Solid-State Relay Handbook

with Applications

Solid-State Relay Handbook

with Applications

ANTHONY BISHOP

Howard W. Sams & Co.
A Division of Macmillan, Inc.
4300 West 62nd Street, Indianapolis, IN 46268 USA

To Robert

FIRST EDITION
FIRST PRINTING—1986

International Standard Book Number: 0-672-22475-5
Library of Congress Catalog Card Number: 86-60028

Edited by: *Louis J. Keglovits*
Designed by: *T. R. Emrick*
Illustrated by: *Wm. D. Basham*
Cover Art by: *Joe LaMantia*

Printed in the United States of America

Contents

Foreword .... *viii*

1 Introduction to Solid-State Relays *1*
1.1 What Is an SSR? *1*
1.2 Hybrid SSR Versions *1*
1.3 Where Are SSRs Used? *3*
1.4 Why Use an SSR? *3*
1.5 SSR vs. EMR *5*

2 Coupling Methods *9*
2.1 Optical (Photo) Coupling *9*
2.2 Transformer Coupling *13*

3 Output Switching Devices *15*
3.1 DC Switches *16*
3.2 AC Switches *20*
3.3 Triac vs. SCR *24*

4 SSR Operation *27*
4.1 DC Inputs *27*
4.2 AC Inputs *28*
4.3 The Coupler *30*
4.4 Hysteresis *30*
4.5 DC SSR *31*
4.6 Zero Switching *32*
4.7 AC SSR *33*
4.8 Additional SSR Circuits *34*

5 SSR Characteristics *41*
5.1 Input Parameters *43*
5.2 Output Parameters *45*

5.3 General Parameters 46
5.4 Selecting the Proper SSR 47

6 Driving the SSR 51
6.1 TTL Drive Methods 54
6.2 IC and Other Drive Sources 55
6.3 Leakage from the Drive Source 56
6.4 A Multifunction Driver 57

7 Thermal Considerations 59
7.1 Thermal Calculations 60
7.2 Manufacturers' Ratings 64
7.3 Heat Sinking 66

8 Surge Ratings Versus High Inrush Current
Loads 71
8.1 Surge Ratings 72
8.2 Inductive Loads 75
8.3 Transformer Switching 76
8.4 Switching Techniques 83
8.5 Motor Switching 84
8.6 Lamp Switching 89

9 Protective Measures 93
9.1 Noise Susceptibility 93
9.2 dv/dt (Rate Effect) 95
9.3 Snubbers 98
9.4 Suppressors 102
9.5 Diodes and Zeners 103
9.6 MOVs 106
9.7 Fuses 112

10 Input/Output Interface Modules for
Microcomputers 121
10.1 What is an I/O Module? 121
10.2 I/O Module Types 123
10.3 Input Modules 123
10.4 Output Modules 126
10.5 Buffered Output Modules 126
10.6 I/O System Concepts 128

11 SSR Applications 131
11.1 Complementary Power Switching 132
11.2 Latching SSR 132
11.3 Latching SSR with Short-Circuit Protection 133
11.4 Fast Response AC/DC SSR 134
11.5 SSR/EMR Hybrid Relay 135

11.6   Motor Starter Switch *135*
11.7   Reversing Motor Drive for Split Phase Motors *136*
11.8   Switching Highly Inductive Loads *137*
11.9   Over/Under Voltage Sensor *138*
11.10 Functional Three-Phase Switch for Three-Wire System *140*
11.11 Phase-Sequence Detector *140*
11.12 Driving High Powered Thyristor and Bipolar Devices *141*
11.13 DPDT Switch from Single Transistor Source *142*
11.14 Special Function Switching with Bounce Suppression *144*
11.15 Temperature Controller *145*
11.16 Phase-Controlled Dimming *146*
11.17 Time Delay/Multivibrator *147*
11.18 Hazardous Applications *148*
11.19 Three-Phase Motor Reversal *150*
11.20 Switching Dual Supplies *152*
11.21 Reversing Motor Drive for DC Motors *154*
11.22 Paralleling SSRs *155*
11.23 Transformer Tap Switching *156*

12   Testing the SSR ***159***
12.1   Dielectric Strength (Isolation) *160*
12.2   Turn-On Voltage, Turn-Off Voltage, and Input Current *161*
12.3   Turn-On and Turn-Off Times *162*
12.4   Off-State Leakage *163*
12.5   Transient Overvoltage (Nonrepetitive Peak Voltage *165*
12.6   Peak Surge Current (Nonrepetitive) *165*
12.7   Zero Voltage Turn-On *166*
12.8   Peak Repetitive Turn-On Voltage and On-State Voltage Drop *168*
12.9   Operating Voltage and Load Current Ranges *170*
12.10 dv/dt (Rate Effect) *171*
12.11 Failure Modes *172*

13   Manufacturers and Cross-Reference Diagrams ***175***
13.1   Available SSR Package Styles *176*
13.2   Directory of SSR Manufacturers *185*

14   Glossary of Commonly Used SSR Industry Terms ***191***

Appendixes

A   Additional Reference Material ***203***

B   Useful Data ***209***

Index ***219***

Foreword

It was evident with the emergence of semiconductor technology in the 1950s that a new era of electrical power switching had begun. With the development of power thyristors, the designer gained a new dimension in switching performance (e.g., low noise, long life, fast switching speed, high reliability, and immunity to shock and vibration) hitherto unattainable with conventional switching methods.

For decades, the electromechanical relay (EMR) had been the primary component used for switching electrically isolated circuits. However, with advancing technology, designers required compatibility with their logic circuits and performance matching that of semiconductors. Subsequently, in the early 1970s the solid-state relay (SSR) appeared on the market, incorporating semiconductor switching circuitry that had for years been the exclusive domain of the circuit designer. These four-terminal, prepackaged assemblies were functionally similar to the EMRs, including isolation between the input circuit (coil) and the output semiconductors (contacts), but there the similarity ended.

Since their introduction, SSRs have found acceptance in a broad number of applications, particularly in such areas as microprocessor-based equipment for industrial machine and process controls. Although the SSR made available the advantages of semiconductor switching in a convenient form, it has since been realized that they have certain characteristics which require special attention to ensure reliable operation. In selecting the proper SSR for the job, consideration should be given to load-related parameters, voltage/current transients, mounting methods, and thermal conditions, among others.

The purpose of this handbook is to give the reader insight into solid-state relays—what they are, how they work, and how to select, specify, test, and generally apply them. Suggested drive and protective methods are included, with application notes to help stimulate further design ideas. It is hoped that this publication will serve as a useful guide and ready reference for those interested in this state-of-the-art component.

1

Introduction
to Solid-State
Relays

1.1 What Is an SSR?

An SSR (solid-state relay) is a mechanically passive version of its older counterpart the EMR (electromechanical relay), providing essentially the same performance, but without the use of moving parts. It is a totally electronic device that depends on the electrical, magnetic, and optical properties of semiconductors and electrical components to accomplish its isolation and relay-switching functions (Fig. 1-1).

The solid-state relay industry, to date, has not been noted for its standardization of relay packages. However, the rectangular package (top center in Fig. 1-1), introduced by Crydom Controls in the early 1970s, has become an industry standard for power switching, with models ranging from 2 to 90 amps. The newer input/output (I/O) modules introduced by Opto-22 (bottom right in Fig. 1-1) have also established a commonality between manufacturers, and in many cases are pin-for-pin replaceable.

1.2 Hybrid SSR Versions

There are available certain relay types that combine the properties of both the SSR and the EMR. They are categorized and described as follows:

Fig. 1-1. *Typical solid-state relay and I/O module packages. (Courtesy IR Crydom)*

1. Hybrid Electromechanical Relay (HEMR)
   Utilizes semiconductors and electronic components for input and drive functions. Isolation is provided, and the output is switched by means of an electromechanical device.
2. Hybrid Solid-State Relay (HSSR)
   An electromechanical device (generally a reed type) providing input, isolation, and drive functions. The output is switched by means of one or more semiconductors.

A third type, which is in effect a parallel combination of an SSR and an EMR, has electronic sequencing components, a photocoupler, and a coil for its input and isolation functions. The output is comprised of a semiconductor switch in parallel with a set of mechanical contacts, each operating in a prescribed sequence (Fig. 11-5). This dual system now becoming available as a prepackaged unit is not yet categorized as a single component.

With time-delay functions added, the electromechanical or solid-state relay becomes a TDEMR or TDSSR. The applications section shows how SSR capabilities may be further extended by means of added logic.

1.3 Where Are SSRs Used?

Since its introduction over 15 years ago, the SSR has invaded most of the application areas of the EMR but has completely replaced it in only a few. Its major growth areas have been associated with the newer computer-related applications, such as the rapidly expanding field of electronic industrial-control systems—specifically, process control, machine-tool control, and energy-management systems that utilize advanced microprocessor-based programmable controllers. The input/output (I/O) modules described in Chapter 10 are a specialized type of SSR designed specifically for computer interface applications.

Because of their low input-control current requirements, SSRs can operate directly from most of the logic systems used in computers today—hence their attractiveness in this area. They can control a wide variety of loads such as motors, lamps, heaters, solenoids, valves, and transformers. The list of applications in which they are used is endless. The following are a few typical examples:

Security Systems	Production Equipment
Fire and Alarm Systems	Test Systems
Dispensing Machines	Vending Machines
Traffic Control	Commercial Laundry
Navigation Equipment	Office Machines
Temperature Controls	Medical Equipment
Instrument Systems	Lighting Displays
Amusement Park Rides	Elevator Controls

1.4 Why Use an SSR?

When properly used, the solid-state relay (SSR) provides a high degree of reliability, long life, and reduced electromagnetic interference (EMI), together with fast response and high vibration resistance, as compared to the electromechanical relay (EMR).

The SSR offers the designer all the inherent advantages of solid-state circuitry, including consistency of operation and a typically longer usable lifetime. This is possible because the SSR has no moving parts to wear out or arcing contacts to deteriorate, which are the primary causes of failure of the EMR. The long-term reliability of solid-state components has been well established, with idealized cyclic (MCBF) rates well into the billions, whereas the useful life of an EMR with a specified 1,000,000-cycle rating could be very short in a fast-switching application (e.g., 280 hours at 1 cps). With virtually no mechanical parts to become detached or to resonate, the ability of the SSR to withstand vibration and shock is also far greater that that of the less rugged EMR.

The input control power requirements of the SSRs are generally low enough to be compatible with most IC logic families (TTL, DTL, HiNIL, etc.) without the need for intermediate buffers or drivers (typically 10 to 80 milliwatts). Control power requirements are in the region of 200 milliwatts for a typical EMR. Some specialized SSR types known as buffered I/O modules are capable of being driven by the lower current of CMOS and NMOS IC logic systems, requiring only 250 microwatts, or less, to operate (Section 10). From the low level signal demands of SSRs, power loads up to 43,000 watts can be controlled with currently available models.

Table 1-1. *Solid-State Relays (SSR)*

ADVANTAGES	DISADVANTAGES
Zero voltage turn-on, Low EMI/RFI	Higher voltage drop (on voltage)
Long life (reliability) $>10^9$ operations	Significant power dissipation—may require heat sink
No contacts—handles high inrush current loads	Off-state leakage—can affect load or be hazardous
No acoustical noise	Cost
Microprocessor compatible	Only SPST easily
Design flexibility	Generally designed for AC or DC loads but not both
Fast response	
No moving parts	Generally can't switch small signals such as audio
No contact bounce	

Table 1-2. *Electromechanical Relays*

ADVANTAGES	DISADVANTAGES
Cost	Slow response
Low contact voltage drop	Random (nonzero) turn-on noise
No heat sink required	Contact wear—short life
No off-state leakage	Poor with high inrush current loads
Multiple contacts	Noisy acoustical
Can switch AC or DC with equal ease (although ratings may not be equal)	More difficult to interface with microprocessors
Higher line power	

Compared to the EMR, the SSR provides a substantial reduction in EMI (electromagnetic interference). Controlled initial turn-on of the SSR at the zero crossing point of the ac line serves to reduce the high inrush currents associated with capacitive loads and lamp loads. The inherent zero current turn-off feature of thyristors used in AC solid-state relays, irrespective of zero voltage switching, provides a dramatic improvement over the arcing contacts of the EMR. With small step changes in power, proportionately lower levels of EMI are generated.

1.5 SSR vs. EMR

With all the aforementioned attributes of the SSR, one might predict the demise of the EMR. Obviously, such is not the case, since the EMR market is still many times greater than that of the SSR. Examination of Tables 1-1 and 1-2, listing the advantages and disadvantages of each, shows that they complement rather than replace each other. The strengths of one are the weaknesses of the other.

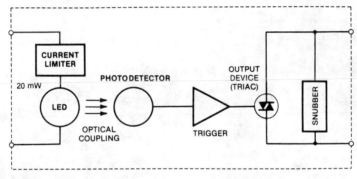

(A) AC solid-state relay (SSR).

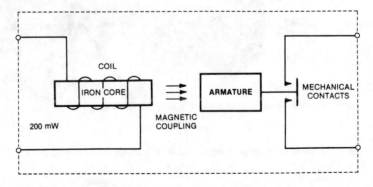

(B) Electromagnetic relay (EMR).

Fig. 1-2. Solid-state relay and electromagnetic configurations.

5

The SSR and EMR are fundamentally similar in that each has an input electrically isolated from the output that controls a load. Fig. 1-2 shows the basic configurations of both the SSR and EMR. In one case, isolation is achieved by photocoupling (Fig. 1-2A) and in the other by means of magnetic coupling (Fig. 1-2B).

Comparing the two relay types, the input circuit of the SSR is functionally equivalent to the coil of the EMR, while the output device of the SSR performs the switching function of the EMR contacts. The SSR is generally limited to a single-pole, single-throw (SPST) device, mainly because of the relatively high cost and volume per pole, whereas, additional poles (contacts) in the EMR add little to the cost or volume, limited only by the magnetic field strength of the coil. Multiple poles in SSRs will become more practical with the increased usage of integrated circuitry, where cost and volume are no longer primary factors.

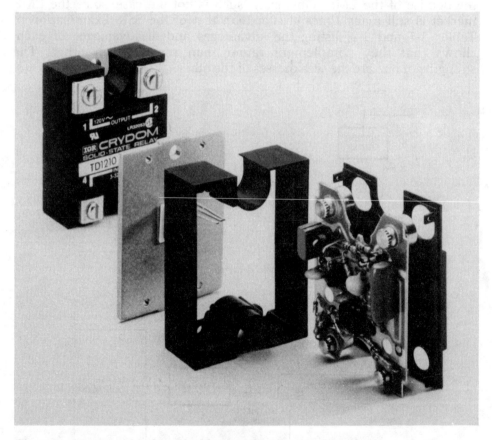

Fig. 1-3. *An open view of a typical power SSR. (Courtesy IR Crydom)*

An open view of a typical power SSR, showing the heat sink and output semiconductor assembly, together with the PC board and terminal assembly loaded with discrete components is depicted in Fig. 1-3. More recent models may contain surface-mount components or a single integrated circuit driver module.

Operating speed of the EMR is dependent on the time it takes for its mechanical mass to react to the application and removal of the magnetic field. Operating speed of the SSR is primarily determined by the switching speed of the output device, typically much faster—microseconds for DC SSRs, compared to milliseconds for EMRs. In most AC SSRs, response time is related to phase angle and frequency of the line, and because of the desired zero voltage/current features, may be deliberately prolonged. In the case of AC input control, the operating speeds of both the EMR and SSR are similarly extended due to phase angle and filtering considerations.

2

Coupling Methods

Generally, input to output isolation is achieved by means of optical coupling or by an oscillator-transformer combination. Other possible methods employing the "Hall effect" or piezoelectric devices have not been largely exploited in commercially available SSRs as yet. The Hall effect coupling method more closely resembles the EMR in that it utilizes an electromagnetic field (coil) to activate a field-sensitive Hall effect semiconductor device. The piezoelectric method, on the other hand, depends on the piezoelectric (crystal) element being physically stressed or vibrated (possibly by electromechanical means) to produce a voltage which, in turn, may be used to drive a semiconductor switch.

Both optical and transformer-coupled methods (shown in Fig. 2-1) provide about the same input to output isolation capability, typically between 1500 volts RMS and 4000 volts RMS. Insulation resistance is in the order of 10^9 ohms measured at 500 volts DC and a capacitive coupling of approximately 8 picofarads. Voltage isolation between input or output to case is usually specified with the same value as input to output isolation.

2.1 Optical (Photo) Coupling

Optocoupling is accomplished in various ways (Fig. 2-2), the more common of which are listed in Table 2-1, together with their relative characteristics in regard to SSR design.

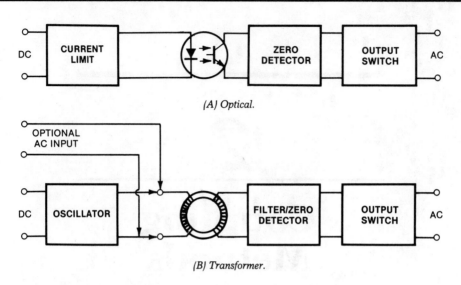

(A) Optical.

(B) Transformer.

Fig. 2-1. *Input isolation achieved with optical or transformer coupling.*

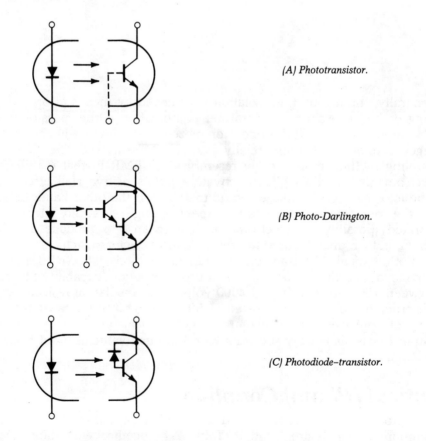

(A) Phototransistor.

(B) Photo-Darlington.

(C) Photodiode–transistor.

Fig. 2-2. *Common LED optocoupling methods.*

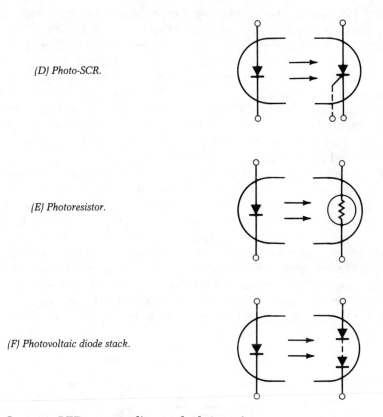

(D) Photo-SCR.

(E) Photoresistor.

(F) Photovoltaic diode stack.

Fig. 2-2. *Common LED optocoupling methods (cont.).*

Table 2-1. *Common Ways of Optocoupling SSRs*

	SPEED	BREAKDOWN VOLTAGE	TEMPERATURE LIMITATIONS	SENSITIVITY
PHOTOTRANSISTOR	Fast	Not a limiting factor in most circuit designs	+100°C max. for LED	Most sensitive
PHOTORESISTOR (CdS)	Slow typ 100 ms	Limited to 400 V for most applications	+75°C max. for cell	Less sensitive than phototransistor
PHOTO-SCR	Fast	Limited to 400 V for most applications	Limited high temp. performance due to noise sensitivity	Relatively insensitive due to added noise suppression circuitry

The phototransistor and photo-SCR types are generally coupled to a gallium arsenide (GaAs) light-emitting diode (LED) source contained within the same package, such as a 6-pin DIP. The minimum current through the LED required to operate the phototransistor in a typical SSR circuit is in the order of 3 milliamperes, and for the photo-SCR it is approximately 8 milliamperes.

The phototransistor, including diode–transistor and Darlington configurations, provides most of the desirable features required for SSR designs and is, therefore, the most commonly used. The photo-SCR is also useful in SSR design, requiring fewer additional components, lower bias requirements, and therefore lower off-state leakage. Where its limitations are acceptable, the photo-SCR type relay may be the more economical, provided the SSR manufacturer has made the right susceptibility–sensitivity trade-offs. This might result in higher SSR input control currents than with the phototransistor types, but would still be within the preferred 16 milliamperes at 5 volts TTL level.

The most commonly used photoresistor or cell type is cadmium sulphide (CdS). This is generally an individual component driven by a separate tungsten or neon lamp; however, it can be configured to operate with less than 16 milliamperes through a closely coupled LED. This type of cell is generally considered to have a limited lifetime and lower reliability, as well as slow speed. The photoresistor, when used as an isolator by adding a light source and a triac (Fig. 2-3), provides the ultimate in SSR simplicity (three components), but it resides at the lower end of the performance spectrum. This type of relay usually serves the appliance market, where a simple isolated switch is all that is required and speed is not important.

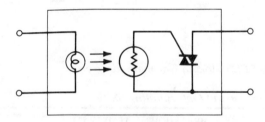

Fig. 2-3. *Simple SSR using a photoresistor driven triac.*

With all the ever increasing specification demands for higher isolation voltages and greater creepage and clearance distances while maintaining close coupling, the "light pipe" (fiber optic) method of coupling the light source to the photodetector will no doubt increase in popularity. Various "lensing"systems are being experimented with in the newer IC relay designs to improve coupling efficiency over that obtainable in commercially available photocouplers.

In addition to higher isolation voltages, the advance in photocoupling technology that is most likely to influence future SSR design is in the area of photovoltaic coupling, whereby usable electrical energy traverses the gap to provide some measure of direct drive (bias) current. This technique permits designs with the advantages of both optical and transformer-coupling methods, e.g., low EMI, low dissipation (DC), and low off-state leakage.

To drive an SSR output, conventional photocouplers require a source of power independent to that of the LED. In a photovoltaic coupling system, the light from the LED produces current in the photovoltaic cell(s) which may obviate or augment the independent power source. At present, the greatest utilization of this coupling technique appears to be in the form of photosensitive multiple diode cell arrays that not only isolate and enhance the transfer of power, but provide the necessary higher bias voltage for driving field effect (FET) devices. Some low level SSRs (less than 400 milliamperes) that utilize these techniques are presently available, which, in a bi-directional form (AC/DC), may eventually emerge as a practical replacement for the electromechanical reed relay.

2.2 Transformer Coupling

While optocoupling is the main isolation system used by solid-state relay manufacturers, transformer coupling also has its merits and is used where best advantage can be taken of the direct control signal drive feature. This type of coupling usually involves a ferrite-cored toroidal transformer with low-loss, high-frequency characteristics. The DC input is chopped by an oscillator operating anywhere between 50 kilohertz and 10 megahertz, while the transformer output is rectified and applied as drive power to the output semiconductors. At the higher oscillator frequencies, junction capacitance in the driven output stage usually provides sufficient filtering to complete the conversion to a steady-state DC bias.

Because bias currents are provided by the control signal rather than by leakage through the load, as in the case of photocoupling, transformer coupling provides a high input/output gain efficiency and also permits lower output off-state leakage, typically less than 1 milliampere without a snubber.

Transformer coupling can be further utilized in DC SSRs where the output transistor is driven into saturation. This results in a lower forward voltage drop (0.2 volt is typical) and less power loss than can be achieved with a four terminal optocoupler system (typically 1.2 volts). With either system, when a Darlington-type output stage is added for higher gain the saturating capability is lost, unless an additional fifth terminal is added accessing the power supply side of the load (Section 3.1).

Since most SSR applications require DC control, the greatest disadvantage of transformer coupling is the need for an oscillator in the primary and, to a lesser extent, a rectifier/filter in the secondary. The oscillator can generate a very low level EMI noise that may be objectionable in some applications. The most practical drive method requiring no oscillator would be directly from a sinusoidal AC source. However, this configuration is rarely found in a production solid-state relay, mainly because the transformer size would be large at low commercial line frequencies (60–400 hertz).

Where a high-frequency clock signal already exists at the driving source (Fig. 2-4) direct transformer drive could be used to advantage. This is not uncommon and could effect a considerable cost savings in a custom built system requiring multiple AC switched points. In this case, the control signal on-off command would appear at the driving output gate as the presence or absence of the clock signal, rather than as a logic level change. Additional components for suppression of the back wave or for integration of the AC signal may be required, but the transformer providing isolation would be extremely small. A further advantage of such a system would be the ease of zero-voltage turn-on synchronization of many switches from a single source.

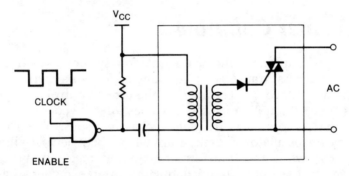

Fig. 2-4. *Simple transformer driven AC switch.*

A possible limitation of transformer coupling is that most SSR circuits cannot be readily adapted to the "normally closed" (form B) configuration. The reason is that the input control voltage usually supplies the output drive power, without which the relay cannot function. This could also be true of a photovoltaic system, or of any output circuit that depends on the control signal for its drive power. Most photocoupled SSRs receive their drive power through the load from the output power source and therefore can be configured either way.

3

Output
Switching
Devices

The AC or DC designation of an SSR generally describes its output switching capability as opposed to its input control voltage requirements, which can also be AC or DC. Most SSRs are configured with a "one form A" (SPST-NO) output, mainly because a multipole arrangement would offer little in economy, not having the commonality of components, such as the armature and coil, found in EMRs. Furthermore, the thermal dictates of the higher powered output switches (typically 1 watt per ampere) require large heat-sinking areas and make condensed packaging impractical. In most cases, the multipole logic functions performed by EMRs are more easily implemented in the IC logic, or at the control source in SSR applications.

AC switching can be accomplished with either transistors or thyristors in bridge or inverse-parallel configurations. Transistors have the added capability of switching DC as well as AC, and they are not plagued by the regenerative rate effects (dv/dt) associated with thyristors. However, thyristors are favored for AC because of their higher current, surge, and voltage capabilities; power transistors are presently best suited to DC switching.

Recent advances in the power switching capability of MOSFETs (metal oxide field-effect transistors) will make these devices eventual candidates for AC and DC power switching in SSRs. Low-power types using MOSFET technology are already on the market. One of the problems, the higher voltage gate drive requirement, has been overcome by series combinations of photovoltaic diodes in the FET gate circuit. Initially, the new smaller units will probably be put into service as drivers for the conventional power output devices.

3.1 DC Switches

The output of a DC SSR is usually a bipolar power transistor, with the emitter and collector connected to output terminals. Fig. 3-1 illustrates the schematic and structure of the two bipolar transistor types, PNP and NPN, the choice of which is primarily a matter of economics, since relay isolation makes it impossible to tell the difference externally. Current flow within the transistors is described by the expression:

$$I_C = I_E - I_B$$

Referring to Fig. 3-2, a family of curves is shown indicating the relationship between base current I_B and collector current I_C. Collector current increases as base current increases along a load line between points "A" and "B," defined as the active region and determined by the load resistance. In switching devices such as SSRs, this region is traversed very quickly (typically less than 10 microseconds), as the drive current from

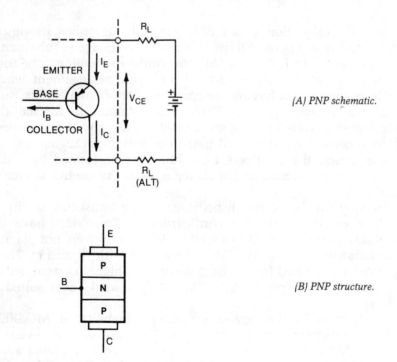

(A) PNP schematic.

(B) PNP structure.

Fig. 3-1. *PNP and NPN transistor types.*

the preceding stage is either at I_{B0} for the off state, or in excess of I_{B6} for the on state. The transition is usually hastened by built-in positive feedback or hysteresis, which also prevents "hang-up" and possible destruction in the high dissipation (active) region caused by the slow transition of an input signal.

The ratio of base current to collector current is the gain or amplification factor of the transistor:

$$I_B : I_C = \text{Gain}$$

In DC SSRs, the degree of amplification is directly related to the small available photocoupler current. As a result, the higher the output current rating, the more stages of gain required. As long as polarity is observed, the load can be switched in series with either of the relay output terminals, as is the case for AC SSRs. This is true for any two-terminal isolated switching device. However, there are three-terminal DC output configurations where the load side of the power supply is connected to a separate terminal on the SSR, as shown in Fig. 3-3. The purpose of the third terminal may be to provide entry for additional internal power, or full

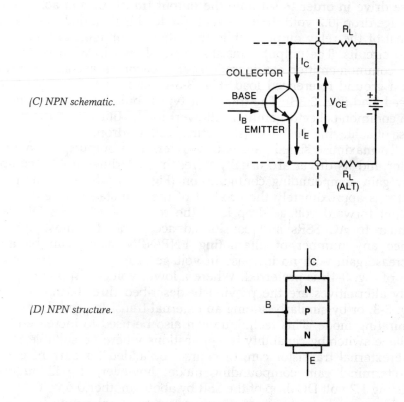

(C) NPN schematic.

(D) NPN structure.

Fig. 3-1. *PNP and NPN transistor types (cont.).*

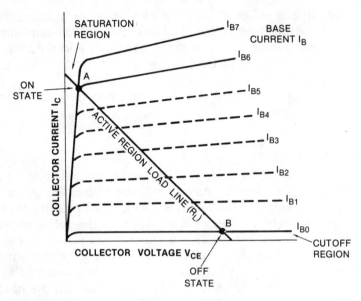

Fig. 3-2. *Transistor voltage–current characteristic curves.*

base drive in order to saturate the output transistor and achieve a lower voltage drop (0.2 volt). In any event, the load is then dedicated to one terminal of the relay output, while the other is common to both drive and load circuits. The output transistor type, shown in what is described as the common-emitter configuration, also becomes a consideration—PNP for a ground referenced load (Fig. 3-3A), and NPN for a positive referenced load (Fig. 3-3B). The transistor types could be reversed and used in the common-collector (emitter-follower) mode, but would defeat the purpose of achieving the lower (saturating) voltage drop.

To maximize signal gain with two-terminal outputs, the output transistor and its driver are usually wired in a Darlington or a complementary gain compounding configuration (Fig. 3-4) where the amplification factor is approximately the product of the two stages. In either case the output forward voltage drop is in the region of 1.2 volts DC, which is similar to AC SSRs and considered acceptable for most applications. Since any number of alternating PNP/NPN stages can be added to increase gain with no increase in voltage drop, the complementary output of Fig. 3-4B is preferred. Where a lower voltage drop is required, the only alternatives are the previously described three-terminal outputs of Fig. 3-3, or by similarly adding an external transistor and driving it in the saturating mode. This technique can also be used to increase current or voltage switching capability in applications where no suitable SSR exists. The external transistor can, of course, be added for current gain in the two-terminal gain compounding mode; however, it will augment the existing 1.2-volt DC drop of the SSR by about another 0.6 volt.

In summation, the more common two-terminal output has the higher voltage drop of approximately 1.2 volts, but it provides the load flexibil-

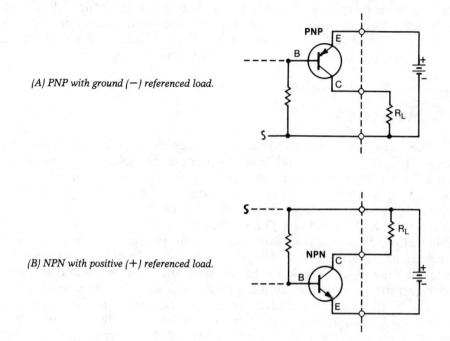

(A) PNP with ground (−) referenced load.

(B) NPN with positive (+) referenced load.

Fig. 3-3. *Three-terminal, DC output, common emitter configurations.*

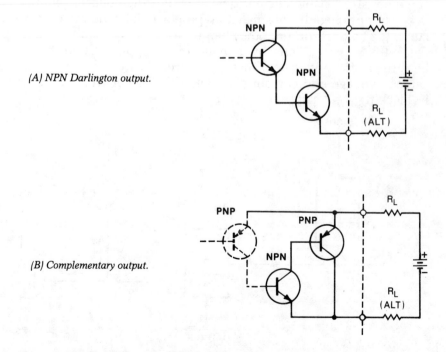

(A) NPN Darlington output.

(B) Complementary output.

Fig. 3-4. *Two-terminal, gain compounding, DC output configurations.*

ity of a true relay. The three-terminal output on the other hand, even with input/output isolation, polarizes the load with respect to the common power supply terminal, but it has the advantage of lower voltage drop (0.2 volt) and, in some cases, lower off-state leakage current.

3.2 AC Switches

The most commonly used output devices in AC SSRs are silicon controlled rectifiers (SCRs) and triacs, generically known as thyristors (so named because of their similarity to the gas discharge thyratrons of the vacuum tube era). Thyristors are a family of semiconductor switches whose bistable state depends on regenerative feedback within a basic four layer PNPN structure. Their attractiveness for SSR use lies in their ability to switch high power loads, with practical values up to 90 amperes and high AC line voltages, up to 480 volts RMS, with less than 50 milliamperes of gate drive. In addition, they can withstand one-cycle peak-current surges in excess of ten times their steady-state ratings.

The SCR is a three-terminal unidirectional device that blocks current in both directions in its off state, and performs much like a rectifier in its on state—thus, a "controlled" rectifier. The SCR is best illustrated by the two-transistor analogy shown in Fig. 3-5. While the transistor can be used as an on/off switch as described in Section 3.1, it is essentially a continuously variable current device where the collector-emitter current flow is controlled by a small but proportional amount of base-emitter current. The SCR, on the other hand, has only two states, "on" or "off." Once it is triggered "on" by a small briefly applied gate signal, it cannot be turned off by its gate. Only with a reversal or reduction of anode to cathode voltage and current below a critical level will the SCR revert to its blocking "off" state.

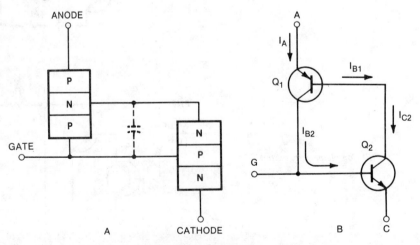

Fig. 3-5. *Two-transistor analogy of SCR operation.*

Fig. 3-5 demonstrates that an SCR function is much like that of PNP and NPN transistors connected in a regenerative feedback loop. Applying a momentary base (gate) signal to Q2 causes it to conduct and supply base current to Q1 which, in turn, supplies a sustaining base current, positive feedback to Q2, holding it on. Conduction is maintained until the "holding" current is reduced and loop gain falls below unity.

The regenerative (latching) characteristic of the thyristor provides its high current and surge capability, but it is also responsible for the thyristor's sensitivity to sharply rising voltages, a less desirable characteristic known as dv/dt, or rate effect. This phenomenon causes inadvertent turn-on, without the benefit of a gate signal. The capacitor shown in A of Fig. 3-5 represents the internal SCR capacitance through which a rising "anode" voltage can inject a turn-on signal into the "gate," resulting in a dv/dt turn-on. In a solid-state relay, this effect is largely controlled by the built-in snubber (RC) network, which limits the rate of rise of the applied voltage. The rate above which turn-on can occur, usually specified on the SSR data sheet as minimum dv/dt, is expressed in terms of volts per microsecond, typically 200 volts per microsecond (Section 9.2). The schematic symbol for the SCR and a typical SCR structure are shown in Figs. 3-6A and B. The structure represents a conventional "edge" or "center" gate fired device commonly used in SSRs. There is also a modified version, used in SSRs for its high stability and superior dv/dt performance, known as a "shorted gate" type. This is where a small portion of the gate-associated P region bypasses the N layer directly to the cathode terminal, providing a resistive path which has the same gate stabilizing effect as that of an external resistor.

The SCR voltage-current characteristic is shown in Fig. 3-6C. Unlike a transistor, the SCR cannot be biased to remain in the transitional zone between its blocking and on states. Once regeneration is initiated, usually by an applied gate signal, the transition is rapid, controlled only by internal positive feedback. The magnitude of the gate turn-on signal depends very little on the anode (collector) current, since conduction is controlled internally; however, it is radically affected by gate impedance, anode voltage, and junction temperature. The relationship is such that a low gate resistance reduces gate sensitivity, while sensitivity increases with higher anode voltages and junction temperatures. Because of these relationships, it can be seen how another turn-on mechanism known as "anode firing" can occur. This is where the forward breakover voltage of the SCR junction is exceeded, and internal "resistive" leakage (rather than "capacitive coupling") from the anode induces a signal into the gate. This method of turn-on is not favored and can be destructive, especially when operating close to the maximum surge current ratings of the SCR.

In its reverse-blocking state, the SCR is similar in characteristics to a reverse-biased rectifier. The maximum reverse-blocking voltage (breakover) is usually specified with the same value as the forward-blocking voltage, but it is much less forgiving if exceeded. In most practical AC

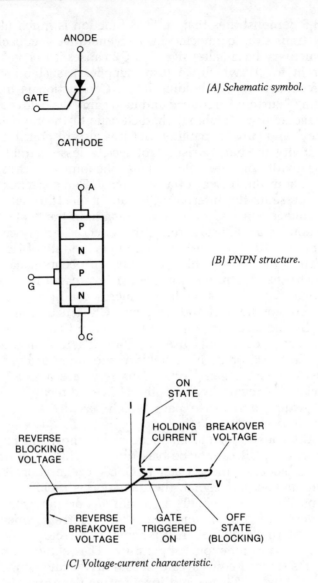

(A) Schematic symbol.

(B) PNPN structure.

(C) Voltage-current characteristic.

Fig. 3-6. *Unidirectional thyristor (SCR).*

circuits, such as a full-wave bridge (Fig. 3-7A) or the inverse-parallel configuration of Fig. 3-7B, the full reverse-blocking capability of an SCR is rarely put to the test. It is protected by the bridge in the former case and by the forward characteristic of the parallel SCR in the latter.

The triac is a three-terminal bidirectional device that blocks current in its off state; but, unlike an SSR, the triac conducts in either direction when triggered on by a single gate signal. As the schematic symbol implies (Fig. 3-8A) the triac is a true AC switch. Its structure (Fig. 3-8B) is essentially that of an inverse-parallel pair of PNPN switches integrated

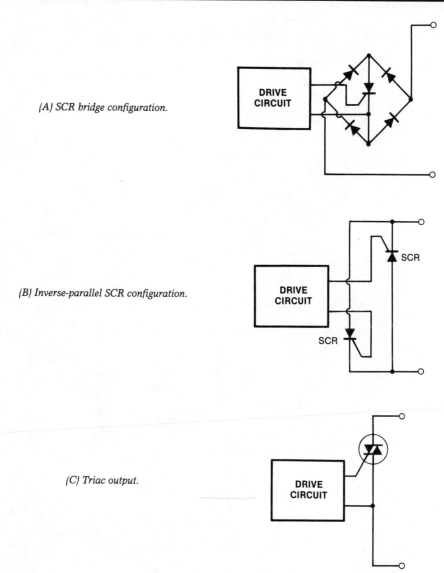

(A) SCR bridge configuration.

(B) Inverse-parallel SCR configuration.

(C) Triac output.

Fig. 3-7. Thyristor output configurations.

into one device. Though the power terminals appear symmetrical, they are designated MT_1 and MT_2 for measurement and biasing purposes. The triac gate is associated with the MT_1 terminal, similar to the gate-cathode relationship of the SCR. Apart from the uniqueness of a single gate controlling oppositely polarized switches with a common signal, the switching characteristics can be likened to those of a pair of SCRs, as can be seen from the voltage-current characteristic of Fig. 3-8C. Even though the two switches are combined into one device, they still exhibit individual characteristics, such as different breakdown voltages, holding currents, and trigger levels.

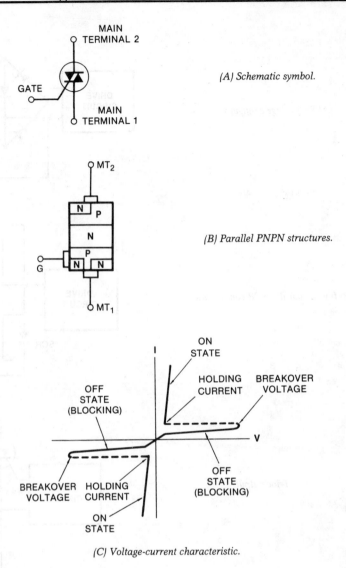

(A) Schematic symbol.

(B) Parallel PNPN structures.

(C) Voltage-current characteristic.

Fig. 3-8. *Bidirectional thyristor (triac).*

3.3 Triac vs. SCR

Prior to the triac, AC power switching was accomplished by SCR bridge circuits such as those shown for comparison in Fig. 3-7. While the rectifier bridge circuit of Fig. 3-7A requires only one SCR, the brief allowable recovery time between half cycles can cause commutation problems. Its biggest drawback is the higher power loss, resulting from the two additional series rectifiers, making it impractical for high current loads.

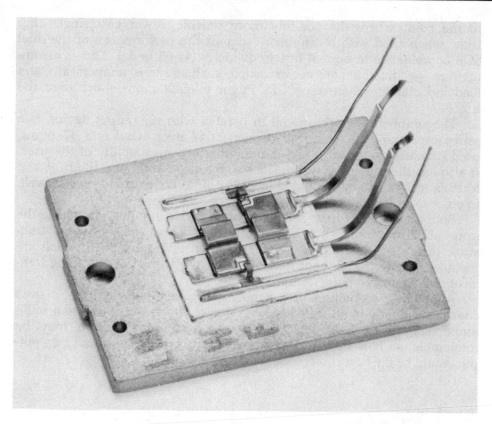

Fig. 3-9. *Power SSR heat sink assembly showing corner gate-fired SCRs in an inverse/parallel configuration. SCR chips are mounted on heat spreading bars which provide more efficient heat transfer through the ceramic substrate to the heat sink base. (Courtesy IR Crydom).*

Though limited as an output switch, the SCR-bridge circuit is used almost universally as a driver stage for the output switch is SSRs, both triac and SCR versions. Used in this way, the higher forward voltage drop is of no consequence and the bridge does a double duty by providing DC to the rest of the driving circuit. The SCR, chosen as a "gate sensitive" type and referred to as the "pilot," does not have the commutating difficulties it might have when switching the load directly. This is because, after triggering the output device, the pilot SCR is shunted off and has the balance of the half cycle to recover.

The inverse-parallel configuration of Fig. 3-7B is the most commonly used for switching the AC line with SCRs. It is relatively free of commutating problems and frequency limitations and is capable of switching thousands of cycles per second due to the full half-cycle recovery time allowed for each element. If size and price were not factors, SCRs would be the logical choice for the SSR AC output switch.

Early triacs were frequency limited and had severe commutation problems. The state of the art in thyristor technology has now advanced to the point where the possibly more reliable, lower cost single chip triac, when used with a "snubber," equals the performance of the dual SCR combination at normal line frequencies (47-63 hertz). This is assuming that specific limits are not exceeded, such as surge, temperature, and load inductance, where the SCRs might tolerate more abuse than the triac.

The snubber network, placed in parallel with the output device, is a series capacitor and resistor (typically 0.047 microfarad and 47 ohms) used primarily to improve the commutating dv/dt capability of the triac. It also has the properties of transient suppression and noise filtering, and for this reason is often included in SCR output configured types as well, in order to equal the performance of the triac versions.

Both the SCR and triac SSRs use similar drive circuits. They have the same noise and transient susceptible areas, such as the "receiver" transistor in the optical coupler and the gate-sensitive pilot SCR, which are the areas where the snubber network is most effective in providing transient protection and noise immunity.

Although the addition of a snubber is generally beneficial, it does contribute greatly to off-state leakage, possibly by as much as 2 to 4 milliamperes. With a light load where the additional leakage may be intolerable, an SSR with inverse-parallel SCRs but no snubber in its output would be the best choice.

4

SSR
Operation

Most SSRs in the higher current ranges are offered with either AC or DC control options. Also, they incorporate some form of current limiting at their input in order to provide a practical operating voltage range. A few models require external current limiting to be provided by the user, but they are usually small low-power types that have little more capability than a photocoupler.

The coil of an EMR is wound to suit its specific control parameters. Similarly, an SSR uses a dropping resistor or a constant-current device to limit current and calibrate its input to suit the desired voltage control range.

4.1 DC Inputs

Figs. 4-1A and B illustrate two typical DC input circuits for controlling current through the photocoupler LED. The low end of the input range is tailored to provide the minimum input current required to operate the SSR, at the specified turn-on (must on) voltage (typically 3 volts DC). The high end of the range is dictated by dissipation in the current-limiting component (typically at 32 volts DC).

Due to its inherently lower dissipation, a constant-current device is capable of providing a broader operating range or occupying less PC board space than a resistor. It may be in the form of a field-effect (FET) as shown, or an IC current regulator. Either type has a minimum

27

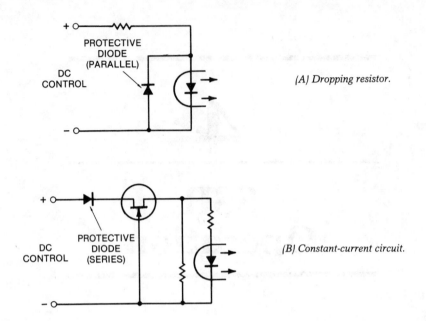

(A) Dropping resistor.

(B) Constant-current circuit.

Fig. 4-1. *Typical DC input circuits.*

operating-current threshold beyond which its impedance increases, thus restricting further current flow with increasing voltage. A disadvantage of the constant-current device, as compared to the resistor, is its relatively low breakdown characteristic (35 volts as opposed to 500 volts typically). Because this device is a semiconductor, this discrete breakdown voltage makes it susceptible to damage by voltage transients.

As a precaution against inadvertent voltage reversal, a series or inverse-parallel diode is usually included in the input circuit (Fig. 4-1). This protection prevents damage to the photocoupler LED and possibly the constant-current device. The series diode permits reversal up to the PIV rating of the diode (typically 600 volts), with negligible reverse-current flow. The parallel diode permits a steady-state reverse voltage equal to the properly applied (forward) voltage, with approximately the same resultant current flow. In this case, the reverse protection is limited by dissipation in the dropping resistor, so the diode or LED will not be damaged by brief voltage transients of a higher magnitude. However, the series diode is favored because it also raises the level of voltage noise immunity by a value equal to its forward voltage drop (0.6 volt approximately).

4.2 AC Inputs

AC input models are usually suitable for both 120 and 240 volt AC line voltages, with a typical operating range of 90 to 280 volts AC and 60 kilohms input impedance. Full-wave rectification is used, followed by capacitive filtering and dropping resistors, as shown in Figs. 4-2A and 4-2B. While both circuits work equally well, the circuit in Fig. 4-2B is favored as being more reliable and fail-safe, since two or more components would have to fail to create an unsafe situation. In the circuit of Fig. 4-2A, a single diode breakdown would place a dead short across the incoming line, thus creating a possible fire hazard.

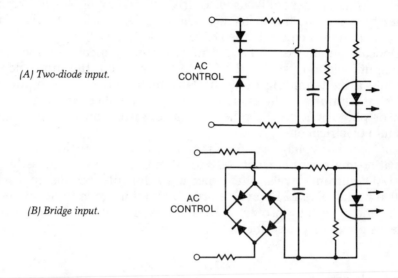

(A) Two-diode input.

AC CONTROL

(B) Bridge input.

AC CONTROL

Fig. 4-2. *Typical AC input circuits.*

Either of the AC input circuits in Fig. 4-2 is also capable of operating from a DC source and therefore might be considered as AC-DC; however, SSR inputs are rarely characterized that way. The circuit of Fig. 4-2B should operate with a DC control range similar to that of the AC (rms) source. On the other hand, the circuit of Fig. 4-2A might have dissipation problems with the input resistors, since they would no longer operate at a 50% duty cycle. In both cases, the SSR would have the uniqueness of operating from a DC signal of either polarity.

A DC controlled SSR may also be converted to AC by the external addition of the bridge and filter circuitry of Fig. 4-2A or B (Section 6.0).

Well designed AC input–output SSRs can operate from separate power sources operating at different frequencies, as long as they are both within the specified limits of voltage, frequency, and isolation. Line frequency for both input and output is typically specified as 47 to 63 hertz, the upper limit of which is not critical for the input control power since the input is rectified and filtered. Also, the lower frequency limit is not critical for the SSR output, since most SSRs can operate down to the DC level. However, the upper frequency limit for an output is less flexible, especially for a triac which has definite frequency limitations related to its ability to commutate off. An SCR output pair is capable of operating at much higher frequencies. However, because of circuit time constants in the drive circuitry, other SSR parameters become the limiting factors (e.g., the zero switching window may be extended and/or turn-on delayed each half cycle with eventual lock-on or lockout).

4.3 The Coupler

DC voltage is generally used to drive the coupling system regardless of the type. Even with transformer coupling, DC is used to drive an oscillator which in turn converts the DC to AC (Section 2.2).

Optical coupling is by far the most common means of achieving input-output isolation (Section 2.1). With this method, the input element is generally a light-emitting diode (LED) which converts the input control power into infrared light energy. This light is collected by a phototransistor or photo-SCR on the other side of the isolation gap and converted back into electrical energy.

The forward-voltage drop of the LED is in the region of 1.2 to 1.8 volts at normal input currents of 2 to 20 milliamperes. At these low nominal LED current levels, life expectancy for the SSR is in excess of 100,000 hours. The LED reverse-breakdown voltage is typically less than 3 volts and is usually protected by a series or (inverse) parallel standard diode, as previously described.

4.4 Hysteresis

Because of the wide variation in photocoupler sensitivities, the minimum voltage to guarantee "off," which is also considered the SSR noise-immunity level, is well below the forward-bias threshold of the LED, typically 1 volt. This threshold can be higher where an additional diode is used in series with the LED. The 2-volt range between the "off" and the maximum operate voltage is an indeterminate state and not largely influenced by hysteresis as in the case for the pickup and dropout of an EMR. The transition is generally made rapidly in either direction, on or off, over a very narrow band, probably less than 0.1 volt, unless hysteresis is deliberately built in.

Hysteresis is where the input voltage required to sustain the output on state is reduced once the transition is made, lowering the turn-off voltage accordingly. Likewise, once the output returns to the off state, the input turn-on voltage is raised back to its initial level. The effect is to speed up the transition and separate the "pickup" and "dropout" control points. In so doing, any adverse threshold effects caused by a slowly ramped-on control signal are minimized.

Hysteresis occurs naturally in an EMR because of its magnetic characteristics. When required in an SSR, it is accomplished by means of a measured amount of positive feedback applied over the output amplifying stages, as illustrated by transistor Q3 in Fig. 4-3. The effect is similar

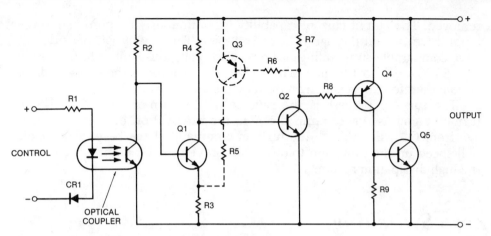

Fig. 4-3. *Optically isolated DC SSR with hysteresis.*

to the regenerative action that occurs during thyristor turn-on described in Section 3.2, except that SSR turn-off capability is only reduced, not completely lost.

The hysteresis characteristic is not generally required in most SSR applications where the thyristors in AC relays have an inherent regenerative action of their own, and the control signals are derived from logic with clearly defined states and rapid transition times, such as TTL. It would be of value however, in high current DC SSRs, where hesitation in the high dissipation, transitional region might be catastrophic to the output transistors, and the resultant "snap action" would reduce or eliminate this possibility.

4.5 DC SSR

The circuit of Fig. 4-3 is an example of a high-current DC SSR incorporating hysteresis. The input control can be DC or rectified and filtered AC. R1 is a current-limiting resistor to protect the LED in the photocoupler, and CR1 provides reverse-voltage protection. With no input applied, the phototransistor in the optocoupler is in its off or high impedance state, and transistor Q1 is permitted to saturate. In this condition, Q2 through Q5 are off, and no power is applied to the load.

When a DC input above the threshold voltage of the LED is applied to the optocoupler, the phototransistor turns on, biasing off Q1. This allows Q2 through Q5 to turn on, and power is applied to the load. Should the turn-on signal be applied in a slowly ramped fashion, Q3 will apply a feedback voltage to the emitter of Q1 which will enhance the turn-off command at its base. This will speed up the turn-on process and thereby hasten the output transistor Q5 through its high-dissipation region.

Unlike an AC SSR which has a latching function, current continues to flow in the drive circuit of a DC SSR, holding it on until the input signal is

31

removed. The output current capability is continuously proportional to the input drive current through the photocoupler. This is why DC SSRs do not have comparable surge ratings. The on-state voltage is similar to that of an AC SSR, 0.8 volt to 1.6 volts, which gives rise to most of the package dissipation; therefore, heat sinking requirements are also similar.

The turn-off process is the reverse of the turn-on (Fig. 4-3). If the turn-off signal is slowly ramped down, the removal of the feedback voltage from Q3 will enhance the turn-on command at the base of Q1. This will speed up the transition to off, again preventing Q5 from hesitating in the high dissipation region.

4.6 *Zero Switching*

Zero voltage turn-on (or zero crossing), as illustrated in Fig. 4-4, is used in most AC SSRs to reduce electromagnetic interference and high inrush currents during initial turn-on. Without zero crossing, the load voltage is applied randomly to the load at any point in the line-voltage cycle. With the zero crossing feature, the line voltage is switched to the load only when it is close to zero, typically specified with a maximum value of ± 15 volts peak. Thus, a very small change in power results, and proportionately lower EMI levels are generated. In addition, high inrush currents such as those associated with incandescent lamps are reduced considerably, thereby extending lamp life. After zero crossing, the "zero" switching voltage, which defines the switching window limits, may also be expressed in terms of phase angle, or time, converted as follows:

Voltage to phase angle (15 volts):

$$\phi = \sin^{-1} \frac{Z \text{ sw. max}}{\text{Line V RMS } (\sqrt{2})}$$

$$= \sin^{-1} \frac{15}{120 \times 1.41}$$

$$= 5°$$

or

Phase angle to time (5°):

$$T = \frac{\frac{1}{2} \text{ cyc. ms}}{\frac{1}{2} \text{ cyc. deg}} \times \phi$$

$$= \frac{8.3}{180} \times 5$$

$$= 0.23 \text{ ms}$$

Zero current turn-off is an inherent characteristic of the thyristors used in AC SSRs, whether zero voltage is employed or not. Once triggered, the thyristor stays on for the balance of the half cycle, until current drops below its "holding" level, where it turns off. For a resistive load, this point is also close to zero voltage, as shown in Fig. 4-4. With an inductive load, the amount of stored energy in the load is a function of the current flowing through it, which in this case is so small that inductive kickback is virtually eliminated. This is probably the most desirable feature of the SSR, when compared to the devastating effects of "arcing" contacts when switching inductive loads with an EMR.

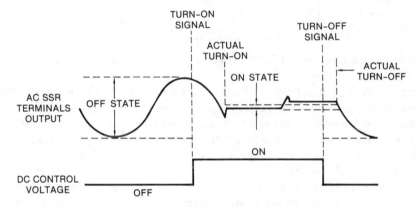

Fig. 4-4. *Control and output terminal voltages for zero voltage turn-on relay.*

4.7 AC SSR

The schematic of Fig. 4-5 illustrates a simplified optically coupled AC SSR circuit, which includes the zero turn-on feature, implemented by the inhibit action of Q1 as described in the following. The input control to the SSR can be DC or rectified and filtered AC. R1 is a current-limiting resistor used to protect the LED portion of the optocoupler, and CR1 provides reverse-voltage protection. With no input applied, the phototransistor in the optocoupler is in its off or high impedance state and transistor Q1 is permitted to saturate. In this condition, the pilot SCR is prevented from firing; thus, the triac is off and no power is applied to the load.

When a DC input greater than the threshold voltage of the LED is applied to the optocoupler, the phototransistor turns on. The values of R2 and R3 are such that Q1 will remain on if the instantaneous line voltage is above zero, thus holding the SSR off until the next zero crossing. When the line voltage rises after passing through zero in either a positive or negative direction, the phototransistor holds Q1 out of saturation long enough for the pilot SCR to trigger, turning on the triac. The triac will

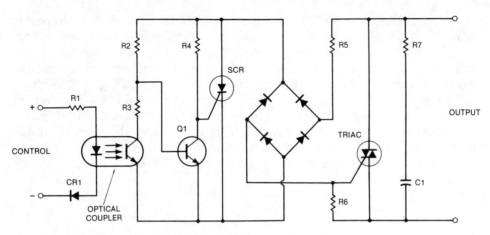

Fig. 4-5. *Optically isolated AC SSR with zero crossing detector.*

0 remain on, being retriggered each half cycle, until the input control is removed and the AC line voltage goes through zero. The result is a continuous sine wave applied to the load, except for a small discontinuity at each zero line crossing, caused by the delay before turn-on. The snubber network of R7 and C1 is used to improve the switching and transient characteristics of the SSR.

The minimum delay for turn-on after zero crossing depends largely on individual circuit design, while the maximum delay is dictated by the zero detector circuit. The initial turn-on point can occur anywhere within these allowable limits, referred to as the "window" or the "notch." Subsequent turn-on points are generally lower and fairly consistent in amplitude, with circuit gain being the primary controlling factor.

Once the output thyristor turns on, the drive circuit is deprived of power by the lower forward-voltage drop of the thyristor, and current ceases to flow. This voltage, which is responsible for most of the package dissipation, varies from device to device and also as a function of the current through it, ranging from 0.8 volt to 1.6 volts. This is why the paralleling of two or more SSRs is difficult, necessitating the use of balancing resistors, etc., to preclude the possibility of current "hogging" (Section 11.18 and 11.22).

4.8 Additional SSR Circuits

The following simplified SSR circuit diagrams have all been used in various production SSR models. Their primary differences lie in input sensitivity, output switching power, and zero switching capability. Other parameters that also seriously affect switching performance are blocking voltage, surge capability, off-state leakage and dv/dt sensitivity. Manufacturers' ratings should be compared, particularly in regard to these and other desired parameters, when making a selection, since schematics are rarely published, and performance characteristics can be significantly different.

In most cases, higher input sensitivity is associated with the phototransistor-coupled types, while photo-SCRs permit lower cost and simpler circuitry. Transformer-coupled types are less common, possibly due to their added drive oscillator and its attendant noise problems.

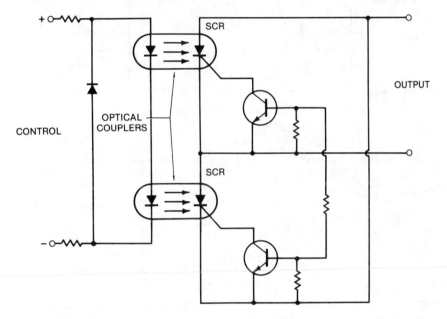

Fig. 4-6. *Low-power AC SSR circuit with dual photo-SCR coupling and zero switching.*

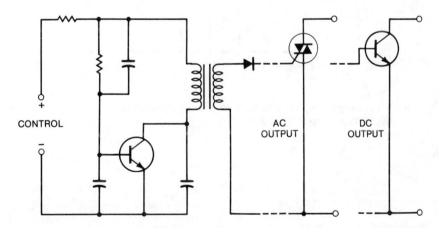

Fig. 4-7. *Low-power AC or DC SSR circuit with transformer coupling, random turn-on, and transistor or triac output.*

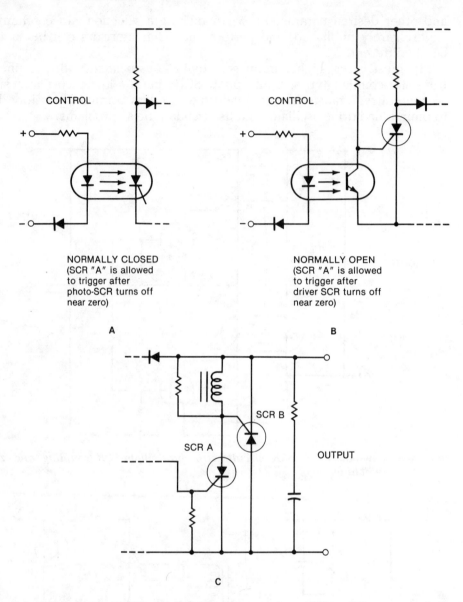

CONTROL

+ o

− o

NORMALLY CLOSED
(SCR "A" is allowed
to trigger after
photo-SCR turns off
near zero)

A

CONTROL

+ o

− o

NORMALLY OPEN
(SCR "A" is allowed
to trigger after
driver SCR turns off
near zero)

B

SCR B

SCR A

OUTPUT

C

Fig. 4-8. *Medium-power SSR circuit with phototransistor-SCR coupling, zero switching, integral cycle SCR output.*

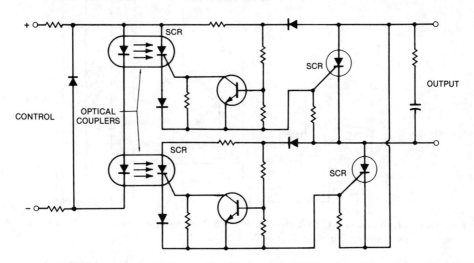

Fig. 4-9. *High-power AC SSR circuit with dual photo-SCR coupling, zero switching and SCR output.*

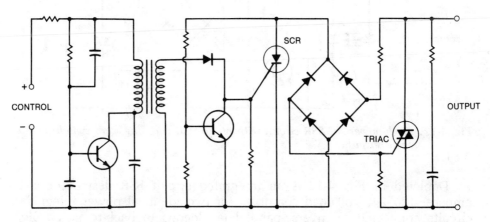

Fig. 4-10. *High-power AC SSR circuit with transformer coupling, zero switching, and triac output.*

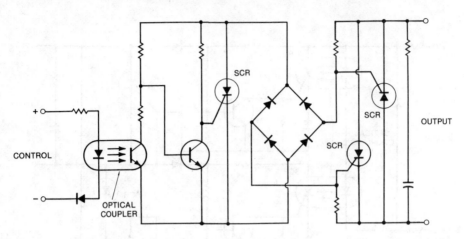

Fig. 4-11. *High-power AC SSR circuit with phototransistor coupling, zero switching (similar to Fig. 4-5), and SCR output.*

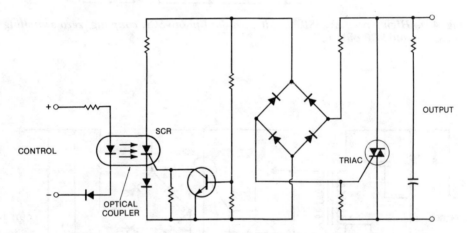

Fig. 4-12. *High-power AC SSR circuit with photo-SCR coupling, zero switching and triac output.*

Depicted in Fig. 4-13 is an integrated circuit SSR design for a 1-ampere, 120–240-volt load comprised of two identical power integrated circuits connected in inverse-parallel (analogous to back-to-back SCRs) for AC control, plus an isolated light-emitting diode (LED) for actuation. Three components form a fully functioning solid-state relay, featuring extremely low zero turn-on and off-state leakage characteristics.

Fig. 4-14 shows a unique 2-pole advanced SSR design that utilizes a bidirectional MOSFET IC for switching low-level signals (AC or DC). Isolating LEDs activate individual photovoltaic generators (shown in cavities) which produce sufficient voltage (5 volts) to turn on the propriety output MOSFET ICs. Intended as a replacement for the electromechanical reed, this relay features low thermal offset (less than 0.2 microvolt), fast response, and long life.

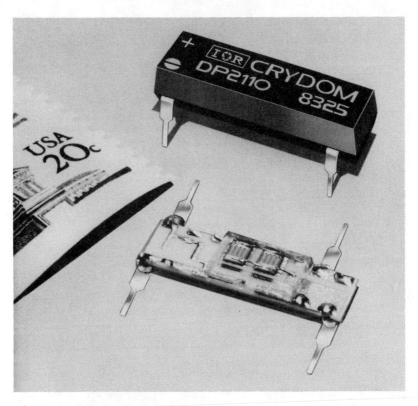

Fig. 4-13. *Advanced integrated circuit SSR design.* *(Courtesy IR Crydom)*

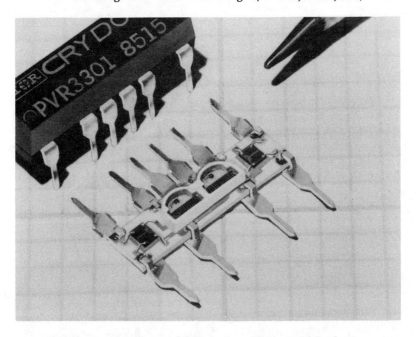

Fig. 4-14. *A unique 2-pole advanced SSR design.* *(Courtesy IR Crydom)*

5

SSR
Characteristics

The first page of an SSR data sheet will usually define the package style, type (AC or DC), and the current-switching range for a particular family of devices. The case styles are many and varied as are the terminals and mounting methods. Some typical examples are shown in Figs. 5-1 and 5-2. The terminal style (screw, fast-on or pin) for each relay is the type most commonly used and best suited to its application range. Fig. 5-3 illustrates graphically the optimum load current application range for the

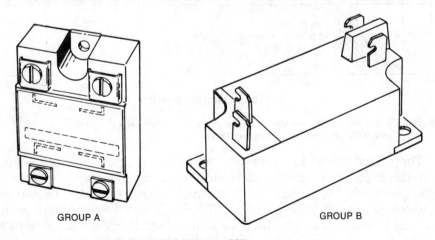

GROUP A   GROUP B

Fig. 5-1. *Typical chassis (heat sink) mount SSRs.*

41

groups of relays shown in Figs. 5-1 and 5-2. "Application effectiveness" is in terms of package suitability versus load current. Upper limits for each category are shown, which with appropriate heat sinking (where applicable) are still constrained by ambient temperature.

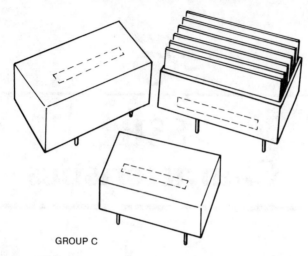

GROUP C

Fig. 5-2. *Typical PC board mount SSRs.*

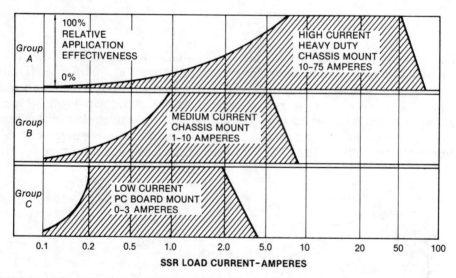

Fig. 5-3. *The SSR types in Figs. 5-1 and 5-2 are shown categorized in terms of optimum current range of application.*

The specifications for solid-state relays are necessarily more complex than those for their electromechanical counterpart, mainly because of some peculiarly "semiconductor" parameters that must be defined (and understood) to avert misapplication. This has resulted in an SSR data sheet that had grown into a hybridized version of both semiconductor and EMR specification sheets, with a few nearly created terms thrown in.

Each style of relay generally has a set of specifications categorized by model number. The specifications are variations of load current, voltage, and input conditions, together with a set of basic parameters common to all models. Table 5-1 shows the characteristics for typical 2, 8, and 40 ampere devices from each of the three groups discussed, listed together for comparison. Sections 5.1 through 5.3 give further definitions for each of the parameters listed. The SSR output configurations are considered to be "single pole, single throw, normally open" (SPST-NO) unless otherwise stated.

5.1 Input Parameters

Control Voltage Range   The range of voltages which, when applied across the input terminals, will maintain an "on" condition across the output terminals. This parameter, inherited from EMR coil specifications, is also used to define (possibly more appropriately) the absolute limits of input voltage that the SSR can withstand without damage (e.g., -32 V DC to $+32$ V DC). In the example of Table 5-1, the lower limit is stated separately as "Max. Reverse Voltage."

Maximum Turn-On Voltage   The voltage applied to the input at or above which the output is guaranteed to be in the on-state. Also known as "must operate" or "pickup" in EMR parlance.

Minimum Turn-Off Voltage   The voltage applied to the input at or below which the output is guaranteed to be in the off-state. Also, known as "must release" or "dropout" and considered to be the SSR "noise immunity" level.

Note: In the band between the turn-on and the turn-off voltages (typically 2 volts), the SSR can be in either state (Section 4.4). For an SSR defined as being "normally closed," the on-off conditions would be reversed.

Maximum Input Current   The maximum current drain on the driving source, usually specified at two points within the control voltage range (output is assumed to be in the on state unless "normally closed"). This defines the input power requirements which can also be given in terms of input impedance at a given voltage.

Maximum Input Current (Off State)   Generally not specified for conventional SSRs, but is usually provided for the more specialized computer interface I/O types (Chapter 10) in the following terms; "Maximum Allowable Input Current for Off State." This parameter determines the tolerable leakage from the control source (rather than the power drain requirements) that will not cause a change in the output state.

Minimum Input Impedance   Minimum impedance at a given voltage which defines input power requirements, as an alternative to, or in addition to, input current.

Table 5-1. *Typical Solid-State Relay Electrical Specifications*

(25°C unless otherwise specified)

OUTPUT CHARACTERISTICS	PACKAGE STYLE			UNITS
	Group A	Group B	Group C	
Operating Voltage Range 47–63 Hz	80–480	20–240	20–240	V_{RMS}
Max. Load Current	40	8	2	A_{RMS}
Min. Load Current	50	20	5	mA_{RMS}
Transient Overvoltage	800	500	500	V peak
Max. Surge Current (Nonrepetitive) 16.6 ms	400	120	55	A peak
Max. Overcurrent (Nonrepetitive) 1 sec	164	30	10	A_{RMS}
Max. On-State Voltage Drop @ Rated Current	2.1	1.6	1.2	V peak
Max. I^2t for Fusing (8.3 ms)	660	60	12.5	A^2s
Thermal Resistance, Junction-to-Case, $R_{\theta JC}(T_J$ Max. = 115 °C(A), 100°C(B))	0.63	3.5		°C/W
Power Dissipation @ Max. Current	60	10	2	W
Max. Zero Voltage Turn-On	75	30	30	V peak
Max. Peak Repetitive Turn-On Voltage	35	15	15	V peak
Max. Off-State Leakage Current @ Rated Voltage (-30°C ≤ T_A ≤ 80°C)	10	4	1	mA_{RMS}
Min. Off-State dv/dt (Static) @ Max. Rated Voltage	200	100	100	V/μs

INPUT CHARACTERISTICS

Control Voltage Range	3–32		3.5–8	V DC
Max. Reverse Voltage	-32		-8	V DC
Max. Turn-On Voltage (-30°C ≤ T_A ≤ 80°C)	3.0		3.5	V DC
Min. Turn-Off Voltage (-30°C ≤ T_A ≤ 80°C)	1.0		1.0	V DC
Min. Input Impedance	1500		225	Ohms
Max. Input Current 5 V DC	4		—	mA DC
28 V DC	20		—	mA DC
3.5 V DC	—		12	mA DC
8 V DC	—		35	mA DC
Max. Turn-On Time (@ 60 Hz)		8.3		ms
Max. Turn-Off Time (@ 60 Hz)		8.3		ms

GENERAL CHARACTERISTICS

Dielectric Strength 50/60 Hz	2500		1500	V_{RMS}
Insulation Resistance @ 500 V DC		10^{10}		Ohms
Max. Capacitance Input/Output		8		pF
Ambient Temperature Range Operating		-30 to 80		°C
Storage		-40 to 100		°C

Maximum Turn-On Time The maximum time between the application of a turn-on control signal and the transition of the output device to its fully conducting (on) state.

Maximum Turn-Off Time The maximum time between the removal of the turn-on control signal and the transition of the output device to its blocking (off) state.

Note: Response times are either assumed or stated as including both propagation and rise or fall (transition) times.

5.2 Output Parameters

Operating Voltage Range The range of voltages applied to the output, over which an SSR will continuously block or switch and otherwise perform as specified. Line frequency is either included or stated separately (AC).

Maximum Load Current The maximum steady-state load current capability of an SSR, which may be further restricted by the thermal dictates of heat sink and ambient temperature conditions (Chapter 7).

Minimum Load Current The minimum load current required by the SSR to perform as specified. Sometimes combined with the maximum load current and given as the "operating current range."

Transient Overvoltage The maximum allowable excursion of the applied voltage that an SSR can withstand without damage or malfunction while maintaining its off state. Transients in excess of this value may turn on the SSR, nondestructively if current conditions are met. The transient period, while not generally specified, can be in the order of several seconds, limited by dissipation in internal bias networks or by capacitor ratings.

Maximum Surge Current (Nonrepetitive) The maximum allowable momentary current flow for a specific time duration, typically one line cycle (16.6 milliseconds) for AC. Usually specified as a peak value and provided with current versus time curves. Relay control may be lost during and immediately following the surge, (Chapter 8).

Maximum Overcurrent (Nonrepetitive) Similar to the above, but typically expressed as an RMS value for a one-second duration.

Note: The duty cycle is not generally specified for either of the above "Nonrepetitive" values, but is considered to be in the order of several seconds and related to the cooling time of the output semiconductor junction.

Maximum On-State Voltage Drop The maximum (peak) voltage that appears across the SSR output terminals at full rate load current. Not to be confused with "Zero Voltage Turn-On" or "Peak Repetitive Turn-On," or used to calculate power dissipation (Section 5.4).

Maximum I²t Maximum nonrepetitive pulse-current capability of the SSR; used for fuse selection. Expressed as "ampere squared seconds" (A^2s) with a stated pulse width, typically between 1 and 8.3 milliseconds.

Thermal Resistance, Junction to Case ($R_{\theta JC}$) Expressed as "degrees celsius per watt" (°C/W), this value defines the temperature gradient between the output semiconductor junction (T_J) and the SSR case (T_C) for any given power dissipation. $R_{\theta JC}$ is necessary for calculating heat-sink values and allowable current and temperature limits (Chapter 7).

Power Dissipation (at Rated Current) The maximum average power dissipation (watts) resulting primarily from the effective voltage drop (power loss) in the output semiconductor. Sometimes provided in the form of curves over the current range.

Maximum Zero Voltage Turn-On The maximum (peak) off-state voltage that appears across the output terminals immediately prior to initial turn-on, following the application of a turn-on control signal. Also referred to as the "notch" which defines the limits of the permissible turn-on window.

Maximum Peak Repetitive Turn-On Voltage The maximum (peak) off-state voltage that appears across the output terminals immediately prior to turn-on at each subsequent half cycle following the initial half cycle, with a turn-on control signal applied. This parameter applies equally to SSRs with or without the "zero turn-on" feature.

Maximum Off-State Leakage Current The maximum (RMS) off-state leakage current conducted through output terminals, with no turn-on control signal applied. Usually specified at maximum rated voltage over the operating temperature range.

Minimum Off-State dv/dt (Static) The rate of rise of applied voltage across the output terminals that the SSR (AC) can withstand without turning on in the absence of a turn-on control signal. Usually expressed as a minimum value at maximum rated voltage in terms of "volts per microsecond" (V/μs).

5.3 General Parameters

As with the EMR, the following three parameters relate to isolation between parts of the SSR, namely input to output, input to case, and output to case. If multiple "contact" types were available, these relationships between the contact sets would need to be defined also.

Dielectric Strength Also referred to as "isolation voltage." Expressed as a voltage (RMS) at 50/60 hertz, that the SSR can withstand without breakdown. Considered as a minimum value.

Insulation Resistance   The minimum resistive value (ohms), usually measured at 500 volts DC.

Maximum Capacitance Input to Output   Maximum value of capacitive coupling between control and power output terminals.

Ambient Temperature Range   The surrounding air temperature limits, usually given for both operating and storage conditions. The maximum operating temperature may be further restricted by the thermal dictates of heat sink and dissipation considerations.

5.4 Selecting the Proper SSR

In specifying the SSR for a particular application, the designer must consider the input, output, load, and installation requirements. In many cases the load power will dictate whether the SSR is PC board mountable or will require a heat sink. Generally, at load currents greater than 5 to 7 amperes, a heat sink becomes necessary, regardless of the current rating of the relay. For example, a typical 10-ampere SSR has a free air rating of 6 amperes at 40 degrees Celsius, whereas a 40-ampere SSR in the same package has only an 8-ampere rating. The improvement, due to a slightly lower dissipation, is small because the two relays are thermally similar and have the same case-to-air interface. This illustrates the point that the full capabilities of a power SSR can only be realized with an adequate heat sink to improve this interface and thereby facilitate the removal of heat from the package. The thermal considerations of the SSR and heat sink selection are discussed at length in Chapter 7.

Having selected an SSR with suitable physical dimensions and terminals, the primary SSR parameters of concern are isolation, input drive requirements, output voltage, and output current. Isolation and input parameters are straightforward and are spelled out much the same as those for the EMR. The output parameters, on the other hand, are more numerous and differ from the EMR, requiring some care in the selection of an SSR in some applications. In many cases the choice of output voltage is simply a matter of selecting an SSR with the appropriate nominal voltage rating (e.g., 120, 240 or 480 volts RMS for AC types) and current ratings in excess of maximum load values (derived from curves).

Most manufacturers have similar operating voltage ranges such as 24 to 140 volts RMS for the 120-volt line and 48 to 280 volts RMS for the 240-volt line; however, the transient overvoltage or blocking ratings may vary by as much as 200 volts. This could make the difference between a reliable installation with an adequate overvoltage safety margin and one vulnerable to destruction by transient voltage spikes. Furthermore, if the transient overvoltage rating is too low, it may even preclude the use of a transient suppressor for protection against such spikes. The range between the SSR peak operating and blocking voltages may be inadequate for the proper function of the suppressor (Sections 9.4 and 9.6).

Some manufacturers suggest that their products will "tolerate" over-voltage transients by "turning on" (firing) nondestructively for the duration of the half cycle in which the transient occurred. This may well be true if normal current conditions are met. However, thus far, this parameter has been poorly defined and therefore should not be trusted without an actual test.

The operating voltage range of 24 to 140 volts RMS for an SSR seems more than adequate, until actual operation at 24 volts RMS is called for. Although it is optimized for the 120-volt line, the SSR will operate at 24 volts RMS and possibly even lower, but again, caution is the keyword. At low line voltages there are two parameters that become significant from the power loss standpoint—on-state voltage drop and peak repetitive turn-on voltage. The former does not change significantly but becomes a larger proportion of the applied voltage, subtracting an approximate 1.2 volts RMS from the load. The latter does not change in amplitude, but it does change in phase angle and therefore represents a considerable delay before turn-on each half cycle, with a power loss of up to 20% at 24 volts RMS, unless the SSR is optimized for operation at this voltage.

The zero voltage turn-on point, which is usually greater than the peak repetitive turn-on voltage, is not a factor in the power loss at low voltage, since it occurs only once during initial turn-on. Of course, the SSR is not much of a "zero" switch at 24 volts either, since turn-on may occur closer to peak voltage than to zero voltage. With less than 24 volts applied, where the peak of the line may become lower than the maximum "zero voltage turn-on" (notch), the switching window remains open and turn-on is totally random.

The SSR "turn-on" voltage parameters are commonly misunderstood, especially when an attempt is made to measure them with an RMS reading meter, which gives erroneous results. Fig. 5-4 shows the three main parameters, exaggerated to illustrate their relationship to each other, and the AC voltage waveform (viewed across the SSR terminals). It can be seen that a discontinuity occurs at the beginning of each half cycle; to a lesser extent, it may also occur at the end of the half cycle, depending on the output thyristor holding current. When a signal is applied during the initial half cycle, turn-on generally occurs at the earliest possible moment in the next half cycle. This point is close to the peak repetitive turn-on voltage. The latest point at which an initial signal can achieve turn-on is at the maximum zero voltage turn-on value. The region between these two parameters is the permissible switching window.

The peak voltage reached at the beginning of each half cycle prior to conduction may be erroneously read as an on-state voltage. The actual on-state voltage is seen as the DC level that follows, with a slight sinusoidal AC component. This is the conducting period which represents most of the dissipation in the SSR, and it should be defined by the manufacturer to aid in heat sink determination.

Equal in importance to voltage are the load-current conditions. The selection of an SSR sometimes requires more involvement with its actual

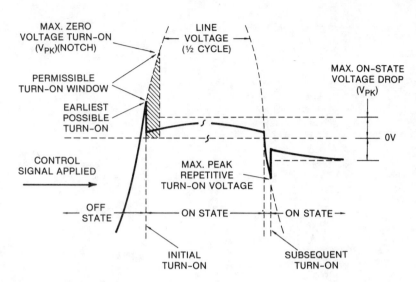

Fig. 5-4. *AC SSR turn-on voltage characteristics.*

application than would be required for an EMR. This is because of the influence the installation and the load might have on its performance, physical size, and the need to provide a heat sink.

The specifications spell out minimum and maximum steady-state load current values, as well as momentary surge current capability. These are expanded upon in the following chapters with detailed explanations of how to apply the SSR in practice and in conjunction with manufacturers' given data. Particular emphasis is given to thermal and surge conditions, which are the areas of most concern.

The preponderance of information on these subjects should not be misconstrued as implying that the user must be fully cognizant of all details in order to successfully apply the SSR. Fortunately, most applications are relatively simple, requiring only a minimum of technical knowledge, most of which is provided by the manufacturer. The following chapters go into more detail and are intended to assist the user in ensuring a reliable installation in a broad area of applications.

6

Driving
the SSR

To activate an SSR output a voltage greater than that specified for maximum turn-on is applied to the input (3 volt DC typical). The off state occurs when zero or less than the minimum turn-off voltage is applied (1 volt DC typical). For an AC input type, the typical values would be 90 volts RMS for on, and 10 volts RMS for off. For an SSR designated as normally closed or form B, the previous on-off conditions would be reversed. Generally, normally open is the accepted, but undesignated, standard for an SSR.

For proper performance and for illustration purposes, ideal supply voltages with good regulation are always assumed, unless otherwise stated. DC is considered as being a steady-state DC voltage of one polarity, and AC is a reasonably well shaped sinusoidal waveform.

Most SSRs are characterized for DC input control signals but may be converted to AC by adding a bridge and filter circuit externally, as shown in Fig. 6-1. The main consideration is that the resultant DC, including ripple, falls within the operating range and input power limits of the SSR. Also, the capacitor (C1) should be sized to raise the DC component above the maximum turn-on voltage value, typically 3 volts DC (i.e., the region between the rectified pulses should not fall below 3 volts). Pulsed (unfiltered) DC causes erratic firing of the output.

A value for dropping resistor R1 can be estimated when the value of input resistance R2 is known, since the ratio is approximately the same as that of the AC source voltage versus the nominal DC input voltage.

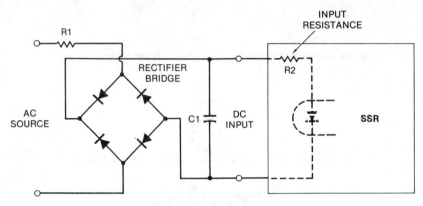

Fig. 6-1. *DC controlled SSR converted to AC control.*

For example (using common values):

$$R1 = \frac{\text{source voltage}}{\text{input voltage}} \times R2$$

$$= \frac{120}{5} \times 1500$$

$$= 36,000 \text{ ohms}$$

For a nominal line voltage of 120 volts AC with a ± 20% tolerance, a rating of 1 watt would be adequate for R1. A suitable value for C1 would be 4.7 microfarads. With this input circuit added, the DC SSR has the properties of operating from control signals of either AC or nonpolarized DC with similar operating ranges.

Because of the input to output isolation, the switch controlling the input to an SSR can be placed in series with either of the two input terminals, assuming polarity is observed (DC). The same flexibility applies to the output side, where the load may also be placed in series with either output terminal. There are a few specialized types, usually with more than two input or output terminals, that have dedicated functions (i.e., V_{cc}, logic input and common).

The activating signal may be derived from mechanical contacts or solid-state devices such as those shown in Fig. 6-2. The minimum supply voltage through the contacts may be equal to the SSR turn-on voltage (3 volts DC typical), whereas the positively or negatively referenced transistors require a minimum supply voltage a few tenths of a volt above the specified turn-on threshold, say 3.5 volts DC. This is because of their approximate 0.2-0.4 volt on-state voltage drop when driven in the grounded-emitter (saturating) mode.

Some other drive configurations, such as the Darlington, with a typical 1.2-volt drop, described in Chapter 3, require still higher minimum supply voltages to compensate for their increased on-state voltages.

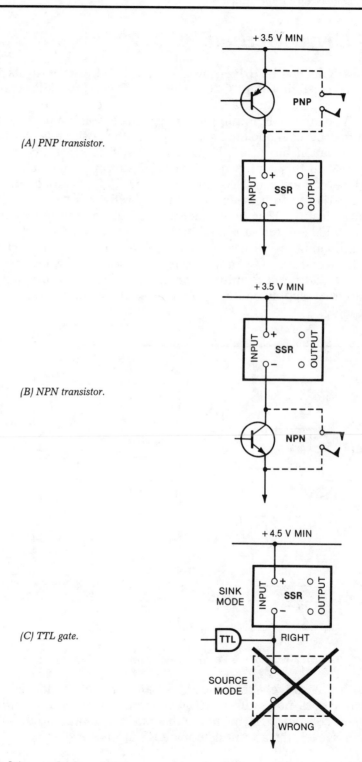

(A) PNP transistor.

(B) NPN transistor.

(C) TTL gate.

Fig. 6-2. *SSR drive methods.*

6.1 TTL Drive Methods

Most SSRs can be driven by a standard TTL gate with its 16-milliampere sink capability (Fig. 6-2C). However, very few SSRs can be driven reliably with the gates' available source current of only 400 microamperes. Also, the SSR minimum voltage threshold requirements are not met in the source mode (i.e., gate output in the positive leg of the SSR).

The relationship of a TTL gate to an SSR is illustrated schematically in Fig. 6-3. In this configuration the SSR supply voltage and the gate V_{cc} should be common and comply with the TTL specified limits of say 5 volts $\pm$ 10%. It can be seen that with a positively referenced SSR and the gate at logical 0, Q2 is operating much like a discrete NPN transistor in the grounded-emitter saturated state. In this mode the gate can sink up to 16 milliamperes with a maximum 0.4-volt drop. Subtracting 0.4 volt from the worst case V_{cc} of 4.5 volts, a minimum of 4.1 volts will appear across the SSR input terminals, which is sufficient to turn on most SSRs. For different supply voltage tolerances, the values would be adjusted accordingly.

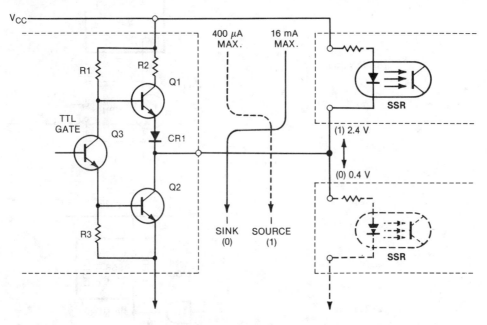

Fig. 6-3. *Typical circuit of a TTL gate driving an SSR.*

With a negatively referenced SSR and the gate at logical (1), Q1 conducts but does not saturate since it is operating as an emitter follower (common collector). In this mode the gate can source up to 400 microamperes; however, the accumulated voltage drops are:

$$R1_{(IR\ DROP)} + Q1_{VBE} + CR1_{VF}$$

54

The sum of these values subtracted from the worst-case V_{cc} results in a minimum output voltage specified as 2.4 volts, which is 0.6 volt below the SSR turn-on threshold (assuming a 3-volt turn-on). Although some SSRs may operate satisfactorily in this mode, it is not recommended that this be done. Both the available current and the minimum voltage are considered inadequate for the typical optically isolated SSR.

It should be noted that the 2.4-volt gate output in the logical 1 state relates only to a negatively referenced load. It does not represent a voltage source to a positively referenced load (SSR), where it would appear to be greater than the off-state voltage. Referring again to Fig. 6-3, Q2 would be off and CR1 is reverse biased, thus presenting essentially an open circuit with virtually zero potential across the SSR.

6.2 IC and Other Drive Sources

Most CMOS and NMOS logic families will not directly interface with SSRs, except for a few specially designed types such as the buffered I/O modules mentioned in Chapter 10. However, a CMOS buffered gate can reliably drive an SSR that has low input power requirements (i.e., >1500 ohms at 5 volts) and is also driven in the sink mode the same as TTL. Fig. 6-4 shows 1/6 of a 4049 (inverting) or a 4050 (noninverting) CMOS hex buffer driving such an SSR with a common 5-volt supply. CMOS can of course operate at higher voltages, but care must be taken to not overstress the gate with excessive dissipation.

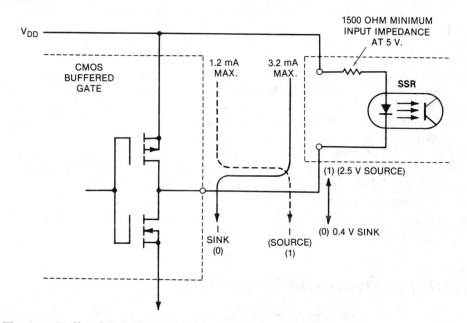

Fig. 6-4. _Buffered CMOS gate driving a high input impedance SSR._

Integrated circuits with open collector outputs are also commonly used to drive SSRs, as in Fig 6-5. The open collector IC has an output transistor without an active (transistor) or passive (resistor) pull-up, and generally has enough power to drive an SSR directly. Open collector outputs can also be logically "ORed" (like discretes), so that the SSR may be controlled by any one of many outputs. Furthermore, the SSR supply voltage does not have to be the same as the IC V_{cc}, provided that one side is common, and the transistor and SSR maximum voltages and currents are not exceeded.

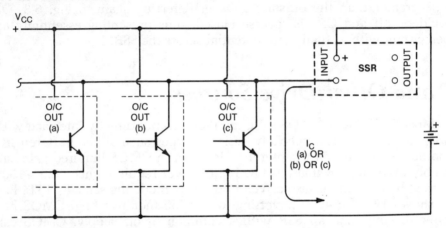

Fig. 6-5. *Open collector IC outputs driving SSR in logically ORed configuration.*

SSRs do not generally require pull-up or shunt resistors for noise reduction or any other functional reason. An open input, if not assigned to a particular logic level, produces an open or off state in the output (unless otherwise designated). Input lines would have to be extremely long and through noisy environments before noise of any significance would appear at the input terminals to cause the SSR to change state.

Some IC devices have "three-state" (tristate) outputs. These have the normal high and low states as described for standard TTL, plus an additional high-impedance state activated by an enable signal. In the high-impedance state, no source or sink current flows, appearing as an open input to a driven SSR. The IC is essentially out of the circuit, thus permitting similar devices to be paralleled and enabled, as desired, without interacting with each other. For example, in this configuration a number of ORed driver stages can be individually polled as to their logic states, by a sequentially applied enable signal. Only the drivers with outputs at logical 0 would activate the SSR.

6.3 Leakage from the Drive Source

The off-state leakage current in the driving semiconductors shown in Figs. 6-2 through 6-5 is insignificant, just a few microamperes, which could not possibly turn on the SSRs. However, the off-state (output) leakage current of any packaged solid-state driving device (e.g., temperature controller, etc.) should first be checked for compatibility with the SSR. One method is to multiply the maximum leakage current (amperes) by the maximum input impedance (ohms) of the SSR. This should result in a voltage that is less than the specified *turn-off* voltage. If it is not, a resistive shunt across the SSR input may be required.

Ironically, one of the more troublesome driving sources is often another SSR (Fig. 6-6). When driving power loads, its off-state leakage is inconsequential, but when it is required to drive another SSR, without a parallel load, it may well be enough to turn on the driven SSR. If, for example, the leakage is specified at 8 milliamperes (max.), it could give rise to a possible 2 volts across an input impedance of say 250 ohms. This would most likely exceed the *turn-off* voltage for the driven SSR and possibly turn it on. A parallel resistance of less than 250 ohms would be required to reduce the resultant voltage below the 1-volt threshold to ensure turn-off.

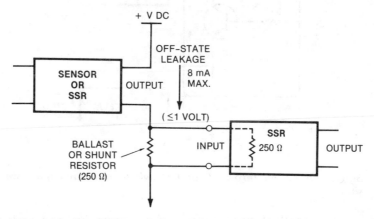

Fig. 6-6. *Off-state leakage of the driver may turn on an SSR.*

It should be noted that the parallel (shunt) resistor will also see the "on" voltage, which may result in an appreciable increase in both dissipation and control current. Sometimes a tungsten filament lamp with just the right characteristic can be used to provide both status indication and act as a shunt that is less wasteful of power. The off resistance of the filament will be far lower than its on resistance. On the other hand, it may be desirable to *increase* control current to satisfy the minimum load requirement of the driver SSR, in which case the parallel resistor serves double duty, both as a shunt and a minimum load.

6.4 A Multifunction Driver

A standard SSR output is off (normally open) with an open input. Since the availability of SSRs with "contact" arrangements other than standard (SPST-N O) is rather limited, it may become necessary for a user to configure his own with external circuitry. A normally closed (N C) function can be implemented with an inverting stage in the drive logic, or a discrete circuit such as that shown in Fig. 6-7. The circuit shows a simple way to achieve a normally open-normally closed pair of SSRs, driven from a positively referenced DC control source. If a normally closed output is required with a completely open input (no power applied), the SSR must be designed that way internally.

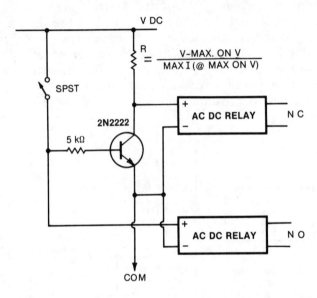

Fig. 6-7. *Single-pole double-throw AC or DC switch from single-pole single-throw source.*

Multiple "contact" arrangements can be configured in the drive logic using individual SSRs; however, caution is necessary because of certain semiconductor characteristics. Fast response and dv/dt can result in "make-before-break" or dual-on conditions, which may be disastrous in reversing circuits or when switching one load between two power supplies. If, for example, the circuit of Fig. 6-7 were to perform such a function (SPDT), additional circuitry would be necessary, such as a time delay between control signals to prevent overlap. Also, since a half cycle short must be assumed due to a possible dv/dt turn-on at power-up, current limiting resistors should be included in series with the SSR outputs. The value should be chosen to limit the instantaneous current to at least the one cycle surge rating of the SSRs. The special requirements of SSRs and the need to custom design multicontact arrangements preclude the high volume manufacture of such devices.

7

Thermal Considerations

A major factor in the use of solid-state relays is the thermal design. It is essential that the user provide an effective means of removing heat from the SSR package. This is because of the relatively high "contact" dissipation, in excess of 1 watt per ampere, compared to milliwatts for the EMR. The importance of using a proper heat sink cannot be overstressed, since it directly affects the maximum usable load current and/or maximum allowable ambient temperature. Lack of attention to this detail can result in improper switching (lockup) or even total destruction of the SSR.

The typical 1.2-volt on-state drop (at maximum current) across the output terminals is responsible for most of the dissipation in both AC and DC SSRs, regardless of operating voltage. At lower currents the voltage falls slightly and creates a nonlinearity in the power dissipation versus current curves, as shown in data sheets.

With SSR loads of less than 5 amperes, cooling by free flowing convection or forced air currents around the basic package is usually sufficient. At higher currents it becomes necessary to effectively expand the exposed radiating surface area by means of a suitable heat sink. This requires that the SSR be firmly mounted to a smooth flat surface on a good heat conductor such as copper or aluminum, plus the use of a thermally conductive compound to improve the interface. Using this technique, the SSR case to heat sink thermal resistance ($R_{\theta CS}$) is reduced to a negligible value of 0.1°C/W (Celsius per watt) or less. This is usually pre-

59

sumed and included in any given thermal data. The simplified thermal model in Fig. 7-1 indicates the basic elements to be considered in the thermal design. The values that are determinable by the user are the case to heat sink interface ($R_{\theta CS}$), as previously mentioned, and the heat sink to ambient interface ($R_{\theta SA}$).

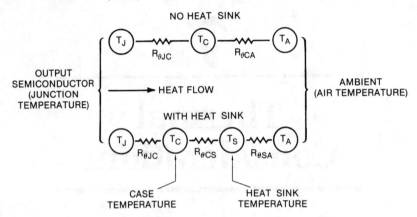

Fig. 7-1. *A simplified thermal model.*

The onus is on the manufacturer to supply the internal thermal values such as maximum junction temperature ($T_J$max.) and the thermal resistance junction to case ($R_{\theta JC}$), or to provide another means of determining thermal performance since these cannot be measured by the user himself. Other important values are power dissipation versus current (specifically at maximum rated current) and thermal resistance of the SSR case to ambient ($R_{\theta CA}$) for free-air performance. Knowing the above values, it is possible to calculate the thermal operational parameters for any application.

7.1 Thermal Calculations

Fig. 7-1 illustrates the thermal relationships between the output semiconductor junction and the surrounding ambient. $T_J - T_A$ is the temperature gradient or drop from junction to ambient, which is the sum of the thermal resistances multiplied by the junction power dissipation (P watts). Hence:

$$T_J - T_A = P\ (R_{\theta JC} + R_{\theta CS} + R_{\theta SA})$$

where

T_J = Junction temperature, °C
T_A = Ambient temperature, °C
P   = Power dissipation ($I_{LOAD} \times E_{DROP}$) Watts
$R_{\theta JC}$ = Thermal resistance, junction to case °C/W
$R_{\theta CS}$ = Thermal resistance, case to sink, °C/W
$R_{\theta SA}$ = Thermal resistance, sink to ambient, °C/W

To use the equation, the maximum junction temperature must be known, typically 100°C, together with the actual power dissipation, say 12 watts for a 10-ampere SSR, assuming a 1.2-volt effective (not actual) voltage drop across the output semiconductor. The power dissipation (P watts) is determined by multiplying the effective voltage drop (E_{DROP}) by the load current (I_{LOAD}). This value is sometimes provided in data sheets in terms of watts at maximum rated current, watts per ampere, or the latter in the form of a curve; the curve is more useful because of the nonlinear relationship between dissipation and current.

Assuming a thermal resistance from junction to case ($R_{\theta JC}$) of say 1.3°C/W and inserting the above typical values into the equation, solutions can be found for unknown parameters, such as maximum load current, maximum operating temperature, and the appropriate heat sink thermal resistance. Where two of these parameters are known, the third can be found as shown in the following examples:

(a) To determine the maximum allowable ambient temperature, for 1°C/W heat sink and 10 ampere load (12 watts):

$$T_J - T_A = P (R_{\theta JC} + R_{\theta CS} + R_{\theta SA})$$
$$= 12 (1.3 + 0.1 + 1.0)$$
$$= 28.8$$

hence,

$$T_A = T_J - 28.8$$
$$= 100 - 28.8$$
$$= 71.2°C$$

(b) To determine required heat sink thermal resistance, for 71.2°C maximum ambient temperature and a 10 ampere load (12 watts):

$$R_{\theta SA} = \frac{T_J - T_A}{P} - (R_{\theta JC} + R_{\theta CS})$$

$$= \frac{100 - 71.2}{12} - (1.3 + 0.1)$$

$$= 1°C/W$$

(c) To determine maximum load current, for 1°C/W heat sink and 71.2°C ambient temperature:

$$P = \frac{T_J - T_A}{R_{\theta JC} + R_{\theta CS} + R_{\theta SA}}$$

$$= \frac{100 - 71.2}{1.3 + 0.1 + 1.0}$$

$$= 12 \text{ watts}$$

hence,

$$I_{LOAD} = \frac{P}{E_{DROP}}$$

$$= \frac{12}{1.2}$$

$$= 10 \text{ amperes}$$

Regardless of whether the SSR is used on a heat sink or the case is cooled by other means, it is possible to confirm proper operating conditions by making a direct case temperature measurement when certain parameters are known. The same basic equation is used except that case temperature (T_C) is substituted for ambient temperature (T_A) and $R_{\theta CS}$ and $R_{\theta SA}$ are deleted. The temperature gradient now becomes $T_J - T_C$ which is the thermal resistance $(R_{\theta JC})$ multiplied by the junction power dissipation (P watts). Hence:

$$T_J - T_C = P \, (R_{\theta JC})$$

Parameter relationships are similar in that solutions can be found for maximum allowable case temperature, maximum load current, and required junction to case $(R_{\theta JC})$ thermal resistance. Again, where two parameters are known, the third can be found as shown in the following examples (using previous values):

(d) To determine maximum allowable case temperature, for $R_{\theta JC} = 1.3°C/W$ and 10 ampere load (12 watts):

$$\begin{aligned} T_J - T_C &= P \, (R_{\theta JC}) \\ &= 12 \times 1.3 \\ &= 15.6 \end{aligned}$$

hence,

$$\begin{aligned} T_C &= T_J - 15.6 \\ &= 100 - 15.6 \\ &= 84.4°C \end{aligned}$$

(e) To determine maximum load current, for $R_{\theta JC} = 1.3°C/W$ and 84.4°C case temperature:

$$P = \frac{T_J - T_C}{R_{\theta JC}}$$

$$= \frac{100 - 84.4}{1.3}$$

$$= 12 \text{ watts}$$

hence,

$$I_{LOAD} = \frac{P}{E_{DROP}}$$

$$= \frac{12}{1.2}$$

$$= 10 \text{ amperes}$$

(f) To determine required thermal resistance $(R_{\theta JC})$, for 84.4°C case temperature and 10 ampere load (12 watts):

$$R_{\theta JC} = \frac{T_J - T_C}{P}$$

$$= \frac{100 - 84.4}{12}$$

$$= 1.3°C/W$$

In examples (a) through (c) SSR operating conditions are determined as they relate to ambient air temperature using a heat sink. Similarly, conditions can be determined for an SSR operating in free air without a heat sink, provided that a value is given for the radiating characteristics of the package $(R_{\theta CA})$. This value is rarely given and when it is, it is more commonly combined with $(R_{\theta JC})$ and stated as $(R_{\theta JA})$. The equation would appear as follows:

$$T_J - T_A = P (R_{\theta JC} + R_{\theta CA})$$

or

$$T_J - T_A = P (R_{\theta JA})$$

where
$R_{\theta CA}$ = Thermal resistance, case to ambient, °C/W
$R_{\theta JA}$ = Thermal resistance, junction to ambient, °C/W

The equation can be used to calculate maximum load current and maximum ambient temperature as before. However, the resultant values are inclined to be less precise due to the many variables that affect the case to air relationship (i.e., positioning, mounting, stacking, air movement, etc.).

Generally, free-air performance is associated with PC board or plug-in SSRs of less than 5 amperes, which have no metallic base to measure. The question is often raised as to where the air temperature is measured. There is no clear cut answer for this. Measurement is made more difficult when the SSRs are closely stacked, each creating a false environment for its neighbor. A rule of thumb is to place a probe (thermocouple) in the horizontal plane approximately 1 inch away from the subject SSR. This technique is reasonably accurate and permits repeatability.

7.2 Manufacturers' Ratings

The free air performance of lower powered SSRs is usually defined on the data sheet by means of a single derating curve, current versus ambient temperature based on the foregoing formulas, which is adequate for most situations. However, a single curve defining performance of a power type on a typical heat sink (1°C/W) is sometimes the only thermal information given. This may provide a means of comparison with another manufacturer's similarly defined part, but it is of little value in determining proper performance in a wide variety of real world applications.

A family of curves such as those shown in Fig. 7-2, while they may be less accurate, provide about the same flexibility as that obtained by calculation. This method of presenting information is particularly useful in that it shows graphically the relationships between all the previously discussed parameters in one, easy to read diagram.

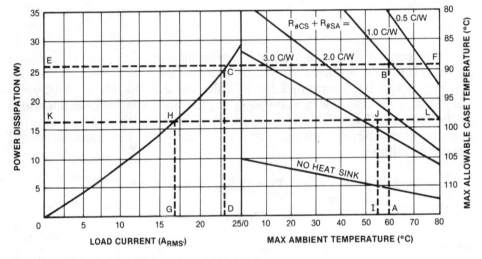

Fig. 7-2. *Thermal derating curves (25 A SSR).*

The left side of Fig. 7-2 shows power dissipation versus load current, a variable factor frequently omitted. The right side is a family of curves used in the selection of the proper heat sink, which in this case may range from 0.5°C/W to 3°C/W. A value is chosen that will maintain the junction temperature (T_J) and the measurable case temperature (T_C) below their maximums for a given ambient temperature. Since the number indicated by the curve includes both the case to heat sink $(R_{\theta CS})$ and heat sink to ambient $(R_{\theta SA})$ thermal resistances, the former must be subtracted from the value selected to determine the required heat sink.

The case to heat sink $(R_{\theta CS})$ parameter is necessarily included in the curves because it is user controlled and can seriously limit the heat sink effectiveness. By following the prescribed technique of mounting the SSR on a smooth heat sink surface and using thermally conductive grease, the value of $R_{\theta CS}$ will be small, 0.1°C/W or less, thus allowing the use of a minimal heat sink. A value for $R_{\theta CS}$ can be found by measuring the differential temperature between case and heat sink $(T_C - T_S)$ and dividing it by the junction dissipation (P watts). Hence:

$$R_{\theta CS} = \frac{T_C - T_S}{P}$$

The broken lines shown on the curves of Fig. 7-2 are to illustrate how the curves are used in conjunction with the following examples:

Example 1

If a 25 ampere rated SSR is mounted on a heat sink which has a thermal resistance of 1°C/W (including $R_{\theta CS}$) and must operate in a maximum ambient of 60°C, the allowable current of 23 amperes may be determined by following the path A,B,C,D. Additional information of power dissipation and maximum allowable case temperature may be found by extending line C,B to points E and F, where the values of 26 watts and 89°C are read.

Example 2

If a current of 17 amperes is required in an ambient of 55°C, the necessary heat sink (plus $R_{\theta CS}$) thermal resistance of 2.7°C/W may be determined by following the path G,H,I,J. Power dissipation and allowable case temperature are found by extending H,J to points K and L, where the values are read as 16 watts and 99°C.

The foregoing information can be used in the selection of a heat sink from manufacturers' dissipation versus thermal resistance curves, such as those shown in Fig. 7-3. The thermal resistance of curve (a) at 16 watts

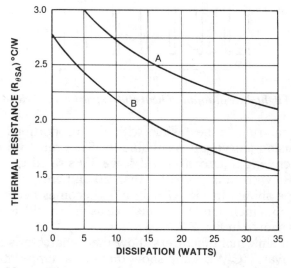

Fig. 7-3. *Typical heat sink characteristics.*

65

is 2.5°C/W. This is better than the required 2.7°C/W in example 2, allowing 0.2°C/W for $R_{\theta CS}$, and is therefore suitable for this application.

Alternatively, in Fig. 7-3 heat sink (b) at 16 watts is 1.9°C/W. Adding 0.1°C/W for $R_{\theta CS}$ and returning to Fig. 7-2, it would allow operation at a maximum ambient of 65°C instead of 55°C.

7.3 Heat Sinking

Confirmation of proper heat sink selection can be achieved by actual temperature measurement on the case (metal base plate) of the SSR. Under worst case conditions the temperature should not exceed the maximum allowable case temperature shown in the right hand vertical scale of Fig. 7-2.

A typical finned section of extruded aluminum heat sink material is shown in outline form in Fig. 7-4. A 2-inch length of this material would approximate the same thermal characteristic as curve (a) in Fig. 7-3; likewise, a 4-inch length would approximate curve (b). This is assuming the heat sink is positioned with the fins in the vertical plane, with an unimpeded air flow.

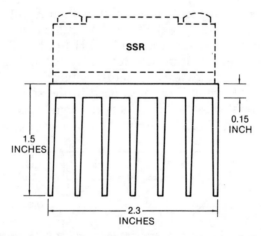

Fig. 7-4. *Typical light duty aluminum heat sink extrusion (end view).*

As a general rule, a heat sink with the proportions of the 2-inch length of extrusion (curve (a)) is suitable for SSRs rated up to 10 amperes while the 4-inch length (curve(b)) will serve SSRs rated up to 20 amperes. For power SSRs with ratings greater than 20 amperes, a heavy duty heat sink of the type shown in Figs. 7-5 and 7-6 becomes necessary. The performance of a 5.5-inch length of this extrusion would approximate the characteristics shown in Fig. 7-7.

Not all heat sink manufacturers show their characteristics in terms of degrees C per watt (°C/W); some show them as a temperature rise above ambient, as shown in Fig. 7-7. In this case, a value for $R_{\theta SA}$ is found by

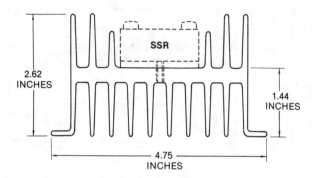

Fig. 7-5. *An end view of a typical heavy duty aluminum heat sink extrusion.*

Fig. 7-6. *A power SSR mounted to the heat sink outlined in Fig. 7-5. (Courtesy IR Crydom)*

dividing power dissipation (watts) into the temperature rise (°C). For example, taking the 60-watt point on the dissipation scale, the free-air curve would indicate a 40-degree rise. Hence:

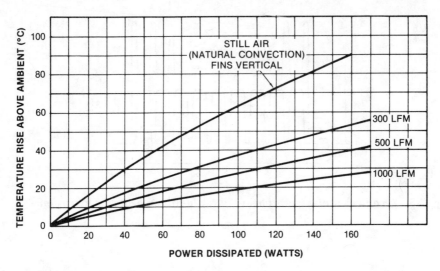

Fig. 7-7. *Typical free-moving air characteristics of a heavy duty heat sink, temperature rise versus power dissipated.*

$$R_{\theta SA} = \frac{T_{Rise}}{P}$$

$$= \frac{40}{60}$$

$$= 0.66°C/W$$

Fig. 7-7 also shows curves that illustrate the effectiveness of forced air cooling on the same heat sink. The improvement is significant when compared to the necessary increase in heat sink size to achieve the same thermal resistance in still air. For example, taking again the 60-watt point on the dissipation scale, the temperature rise at 300 lineal feet per minute (LFM) is only 24°C which, using the foregoing formula, converts to 0.4°C/W. To achieve this value in still air by extending the length of the same extrusion would take approximately a 24-inch length, more than four times as much. Since the efficiency of the extrusion falls off with increased length, extending it further would only slightly improve this figure. An extrusion with a larger cross-sectional area would be necessary to improve performance under the same conditions.

In many applications, the SSR is mounted to a panel or base plate, which may also be more than adequate as a heat sink. By ensuring flatness, using thermal compound, and removing paint to maximize effectiveness, a base plate (SSR) temperature measurement at maximum ambient may be all that is necessary to confirm proper operation as previously mentioned.

The graph in Fig. 7-8 shows the thermal resistance versus surface area of a square 1/8-inch aluminum plate. A temperature rise of 50°C is assumed, which is the typical value at which heat sink manufacturers rate their products. This will give an indication of size and required surface area for a particular thermal resistance. The relay is mounted vertically in the center in still air, allowing natural convection currents to flow.

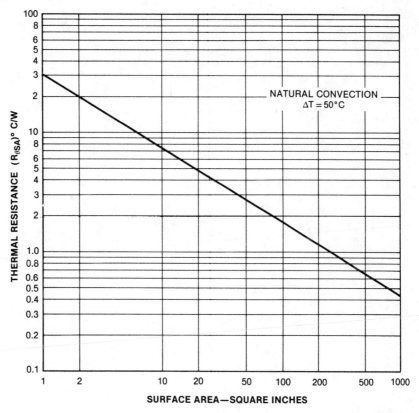

Fig. 7-8. *Thermal resistance versus surface area for a 1/8-inch aluminum plate.*

If an SSR installation does not provide an adequate heat sink, a selection is made from the wide variety of commercial heat sink types that are available. Each configuration has its own unique thermal characteristic and is usually well documented with manufacturers' performance curves and applications data. Much useful thermal information is available from this source, including such details as mounting multiple units on a single heat sink.

8

Surge Ratings Versus High Inrush Current Loads

There are very few completely surgeless SSR loads. Even heating elements, although being resistive, generally exhibit high start-up current due to their positive temperature coefficient. Incandescent lamps are among the worst offenders, with turn-on surges 10 to 15 times their steady-state current ratings. This is a result of their extremely low filament resistance when cold.

Capacitive loads can also be treacherous because of their initial appearance as short circuits. High surge currents can occur while charging, limited only by circuit resistance. Inductive loads on the other hand tend to impede high inrush currents; in fact, inductance is often inserted into a circuit for the express purpose of limiting high fast rising peak currents (e.g., EMI filters, chokes, etc.).

Inductive loads, however, can also give rise to high inrush (magnetizing) currents. Certain solenoids and transformers that are inclined to saturate can create extremely high initial currents, limited only by the DC resistance of their windings. Regulating and variable transformers are particularly troublesome in this regard. Induction motors in their initial stalled state can cause inrush currents 4 to 6 times their nominal running currents.

Inductive loads have traditionally created more problems on turn-off rather than turn-on due to stored energy and "back EMF." The inherent zero current turn-off characteristic of thyristors used in AC SSRs is most beneficial in this regard. Since the amount of stored energy in the load is directly related to current, with turn-off at or near zero current inductive kickback and its attendant ill effects are considerably reduced.

Next to improper heat sinking, surge current is one of the more common causes of SSR failure. Overstress of this type can also seriously impair the life of the SSR. Therefore, in a new application it would be wise to carefully examine the surge characteristic of the load and then select a device from manufacturers' data that can adequately handle the inrush as well as the steady-state condition, while also meeting the lifetime requirements.

In addition to the actual surge ratings given for SSRs, the rate of rise of surge current (di/dt) is also a factor in AC thyristor types; exceeding its value may result in destruction of the device. Since the di/dt value is rarely, if ever, specified, it's difficult to know when this destruct point is reached, except maybe by the unexplained demise of the SSR. As a guide, the amperes-per-microsecond (di/dt) withstand capabilities for the types of SCRs and triacs generally used in SSRs, with specified sustained safe breakdown "anode firing" capabilities, are typically in the order of their single cycle surge ratings. With zero voltage turn-on, sinusoidal waveforms, and typically inductive loads, these critical rates of rising current will rarely be reached (except perhaps with a purely capacitive load). However, SSR manufacturers would be wise in providing such a value for 90-degree firing types and random turn-on (phase-control) types, firing close to peak, where a knowledge of this limit might be useful to the user/designer.

8.1 Surge Ratings

The highest surge current rating of an SSR is typically 10 times the steady-state RMS value, and it is usually given as the maximum nonrepetitive peak current for one line cycle. This characteristic is often illustrated by means of a curve that shows the current declining exponentially to a steady-state condition in about 10 seconds (Fig. 8-1). The curve which represents the loci limits of a peak current of uniform amplitude for a given time is sometimes misconstrued as being the actual shape of the current surge waveform. The waveshape of the surge current contained within these limits is usually considered as sinusoidal (per EIA/NARM standard RS-443), but in some manufacturers' data is stated as being a square pulse. The surge characteristic, which may be preceded and followed by any rated load condition, is developed from thyristor manufacturers' data and is commonly used to define maximum device surge capability.

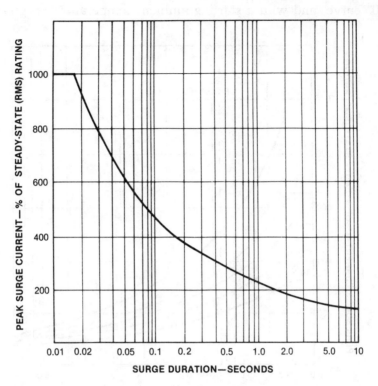

Fig. 8-1. *Typical curve for peak surge versus duration (nonrepetitive).*

While the curve is useful as a not-to-exceed limit, it should be noted that a surge of this magnitude is allowable only 100 times during the SSR lifetime. Furthermore, control of conduction may be momentarily lost, since the output device junction is allowed to exceed its "blocking" temperature, thus causing lock-on, while remaining below its destruct temperature. This means that it may not be possible to turn off the SSR by removal of control power both during and immediately after the surge. The output thyristor must regain its blocking capability and the junction temperature allowed to return to its steady-state value before reapplication of surge current, which may take several seconds.

The preceding cautionary notes would tend to reduce the attractiveness of the high surge capability (1000%) of the AC SSR; however, they apply only to the extreme limits where the SSR should not be designed to operate anyway. When a reasonable surge safety margin is used, conditions rapidly improve, as can be seen from the curves of Fig. 8-2. When supplied by the manufacturer, these surge versus lifetime curves provide far more practical operating information than does the single curve of Fig. 8-1. For example, better than a million (10^6) occurrences are allowed for a surge with a magnitude 500% of steady state. In addition, the input signal remains in full control of the SSR the whole time. The limits at which control is retained are usually stated as being for surges less than the 10^4 curve, and with a starting ambient temperature no greater than 40°C.

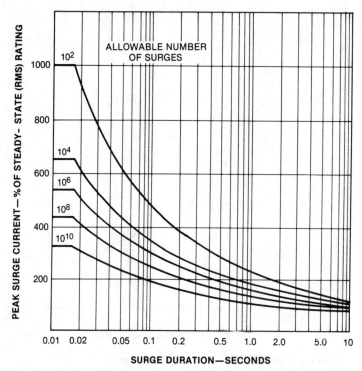

Fig. 8-2. *Curves for surge versus SSR lifetime.*

It can be concluded from the above that proper interpretation of the manufacturers' surge ratings is extremely important for high inrush loads. Underwriters Laboratories take into account the surge characteristics of both the load and the output semiconductors when applying their ratings to SSRs (Section 8.5). Typical motor and lamp ratings can be 25% and 50%, respectively, of their steady-state specified maximums. Conservative ratings such as these are an excellent guide when applying the SSR when only the steady-state load conditions are known.

Generally, DC SSRs do not have an overcurrent surge capability, since the output transistors (nonregenerative) are usually rated for continuous operation at their maximum capacity. The tendency is for the DC SSR to cut off (current limit), thus impeding the flow of excessive current. However, the resultant overdissipation may destroy the relay if the surge is prolonged. If overcurrent carrying capacity is required, as may be the case when designing fuse protection, the SSR could be overspecified (have a higher current capability) to allow the fuse opening current to flow. Examination of both fuse and SSR curves would be necessary for this.

To aid in the proper design of SSR fusing, an I^2t rating is sometimes given. This parameter expressed in ampere-squared seconds is useful since it can relate directly to the published fuse characteristics. It is generally derived from the peak surge (one cycle) output thyristor rating as follows:

$$I^2t = \left(\frac{I^2_{PK(SURGE)}}{2}\right).0083_{(SECONDS)}$$

For example, for a 25-ampere SSR with a 250-ampere one cycle surge rating, the value would be 260 ampere-squared seconds.

8.2 Inductive Loads

High inrush lamp and capacitive loads sometimes include a series inductor such as a choke or transformer. This will tend to limit the initial inrush current, but the combination will primarily be seen by the SSR as an inductive load. While most SSR loads, even lamps, include some inductance, its effect with resistive loads is usually negligible. Only those loads that utilize magnetics to perform their function, such as transformers and chokes, are likely to have any significant influence on SSR operation.

The effect of inductance on an AC circuit is to create a delay in the current waveform causing it to lag behind the voltage waveform. The amount of delay may be described in terms of phase angle, between 0° and 90°, or in terms of power factor, which is the cosine of the phase angle, with values between 0 and 1. Power factor is also the ratio of load resistance to load impedance R/Z. It represents the actual power used by the load compared to the apparent power as seen by the generator. The impedance (Z) is the vector sum of the load resistance (R) and the inductive reactance (X_L), as shown in Fig. 8-3.

For a resistive load the current and voltage are in phase (zero degrees shift) and the power factor is unity (1). For a pure inductor, the current lags the voltage by 90°, and the power factor is zero (0). SSRs are generally specified as being capable of switching inductive loads with minimum power factors ranging between 0.5 and 0.7, which convert to phase angles of 60° and 45°, respectively; this includes most inductive loads.

The majority of SSRs today will operate inductive loads with power factors as low as 0.1, especially if they are switching medium- to high-current loads relative to their rating. However, with lighter inductive loads the possibility that erratic firing (half cycle drop-out) may occur increases. This is caused by insufficient latching current to hold the output thyristor on before the zero switching window closes. The solution is to make current available early in the cycle by means of a resistive shunt, RC network, or a lamp across the load. This will supply minimum holding current and also tend to correct the power factor. A larger snubber (RC) network across the SSR output would also improve this situa-

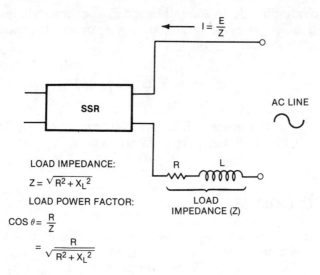

$$I = \frac{E}{Z}$$

AC LINE

SSR

LOAD IMPEDANCE:

$$Z = \sqrt{R^2 + X_L^2}$$

LOAD POWER FACTOR:

$$\cos \theta = \frac{R}{Z}$$

$$= \frac{R}{\sqrt{R^2 + X_L^2}}$$

R   L

LOAD
IMPEDANCE (Z)

Fig. 8-3. *Inductive load switching.*

tion; however, the increased leakage current may create other problems for smaller loads.

When a load is so light that its rating is close to the minimum current rating of the SSR, the off-state leakage may become significant when compared to the load current. The leakage may have a deleterious effect on certain loads such as small solenoids and relays that fail to drop out, or motors that buzz or even continue to run. In the off state, the applied voltage will divide between the "effective" SSR output impedance and the load impedance. For example, a typical 8 milliamperes of SSR leakage will produce a 19.2-volt drop across a 50-milliampere (2400 ohm) load. The solution is again to reduce the load impedance by means of a shunt or parallel impedance, thus reducing this voltage below the drop-out or "off" threshold of the load. (Refer to Section 6.3.)

A saturating inductive load can also cause switching problems with the SSR. The AC impedance of such a load is relatively high under normal conditions. However, when saturation occurs the inductance falls to a very low value, resulting in a fall in impedance close to that of the copper resistance of the coil winding. This can cause several cycles of surge currents in excess of 30 times the steady-state value, which may seriously affect the lifetime, if not cause the destruction, of the SSR (Fig. 8-4).

8.3 Transformer Switching

Extremely high current surges are commonly associated with transformers, especially those with a penchant for saturation. The zero voltage turn-on feature of standard SSRs can increase this possibility and

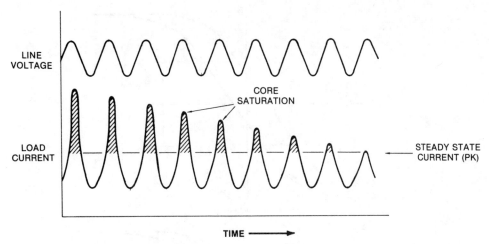

Fig. 8-4. *Current waveform when saturation occurs at turn-on.*

might require that special precautions be taken, or, in extreme cases, that a different switching technique be employed (Section 8.4). The zero current turn-off characteristic of SSRs, while minimizing the problem, will not prevent it.

At the instant of turn-on, transformer current is essentially zero, with the highest peak usually occurring within a half cycle, depending on the line phase angle, load power factor, and magnetic state of the core. When the SSR is energized at the "ideal" phase angle, as dictated by power factor, a maximum back emf is generated that will tend to counter the magnetizing current, thereby reducing or eliminating the surge. However, when switched on at, or near, zero voltage, the back emf is reduced, allowing an increase in magnetizing current that can far exceed the value of nominal peak current. This condition can be further enhanced by residual magnetism in the core, which almost always exists since ferromagnetic core material has a natural tendency to remain magnetized at turn-off.

Fig. 8-5 illustrates three different (single cycle) conditions of a transformer B-H (hysteresis) curve, with flux density shown on the B axis and magnetizing current on the H axis. In Fig. 8-5A, conditions are normal and maximum flux density is well within the limits of saturation. If the transformer is switched off during the negative excursion of magnetizing current, the state of the core would be at point G in Fig. 8-5B. With power reapplied during the first positive half cycle of flux, it can be seen that the peak magnetizing current at point D has now increased along the H axis compared to its previous position.

Referring again to Fig. 8-5A, if the power had been switched off during the positive excursion of magnetizing current, the state of the core would then be at point F in Fig. 8-5C. The reapplication of power during a positive half cycle of flux can now take points D and E far into saturation, thus creating the condition illustrated in Fig. 8-4. Fig. 8-6 illustrates

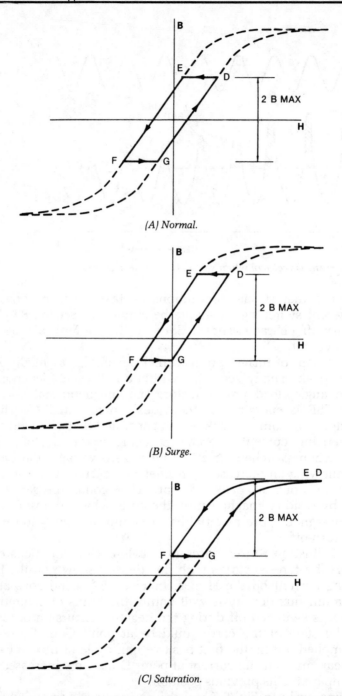

Fig. 8-5. *Transformer hysteresis (B-H) curves illustrating surge conditions.*

the same point, shown in a more graphic sinusoidal-time relationship. This condition of extremely high current can last over several half cycles, diminishing until equilibrium is again reached.

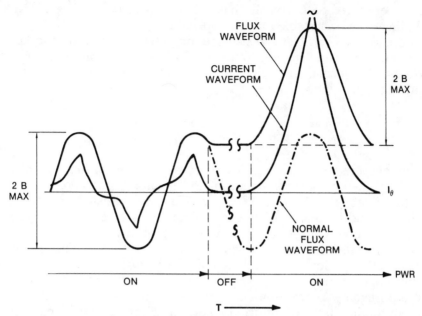

Fig. 8-6. *Flux density versus magnetizing current shown after a power interruption and initial reapplication.*

Saturation can occur with most ferromagnetic (iron) cored transformers, which are often designed for economic reasons to operate very close to the saturation point on their B-H curve. Transformers could be designed with lower flux densities, but at a higher cost and larger size. Some of the transformer types more inclined to saturate are variacs, step-up, step-down, isolation, ferroresonant and regulating types. Those designed for 50-60 Hz operation have a greater safety margin than those designed for 60 Hz operation only.

The magnitude of the current surge depends on the transformer core material, the winding (primary) resistance, and the line impedance. Using the most conservative design approach to ensure that the SSR can handle a surge under saturating conditions, select an SSR that has a one cycle surge rating greater than the worst case peak surge current of the load. For example, a transformer with a primary rating of 120 volts AC and a DC resistance of 0.8 ohm could have a maximum current surge of 211 amperes, determined as follows:

$$I_{surge} = \frac{E_{(LINE)} \, (\sqrt{2})}{R_{(PRIMARY)}}$$

$$= \frac{120 \times 1.41}{0.8}$$

$$= 211 \text{ amperes peak}$$

A 25-ampere rated SSR typically has a 250 ampere peak surge rating and would therefore be suitable for this application.

Alternatively, sufficient resistance may be inserted in the line to reduce the peak surge current to that of the SSR, assuming that it will not interfere with normal operation. The resistance (R_X) should be chosen according to the relationship:

$$\frac{\text{Primary}}{\text{DC resistance}} + \frac{\text{Current limit}}{\text{Resistor }(R_X)} = \frac{E_{PK(LINE)}}{I_{PK(SSR\ SURGE)}}$$

For a 10-ampere SSR with a 100-ampere peak surge rating, operating at 120 volts AC, the total resistance of the primary plus additional resistance (R_X) is about 1.7 ohms. With a primary DC resistance of 0.8 ohm, 0.9 ohm should be added. If the transformer has a 250 volt-ampere rating, the additional voltage drop due to the 0.9 ohm would be (250/120) × 0.9, approximately 1.9 volts, which is probably within circuit tolerances. By this method a lower rated SSR may be allowed without interfering with normal operation.

The foregoing precautions assume a nonrepetitive, initial start-up surge condition that will return to normal within a few cycles. In all probability the actual surge will be far less severe than that suggested, and worthwhile economies may be achieved by actual measurement and examination of the surge waveform. However, without the availability of laboratory facilities, the preceding suggestions, if used, should result in a reliable design.

8.4 Switching Techniques

Some manufacturers recommend random as opposed to zero turn-on for highly inductive loads. Random (or nonzero) turn-on is when the SSR output may be activated at any time during the first half cycle, rather than at the zero crossing point of the line. The theory is that the odds of turning on under worst case conditions are reduced, thereby improving the SSR's chances of survival. Where the theory falls short is that if no other precautions are taken, it may only take that one chance (worst case) turn-on to destroy the device, thus only delaying the inevitable.

Random turn-on does have other virtues. It allows the SSR to be phase-angle controlled by means of an external signal, assuming that the on-off response times permit. It may also be the solution in situations where a highly inductive load is unable to supply sufficient latching current to achieve turn-on of the SSR before the zero switching window closes, causing dropout (Section 8.2). A more serious side effect of this condition is the possibility of the SSR continuously "half-waving" in a marginal situation. The resultant pulsed DC applied to a transformer primary can cause devastating saturation currents to flow *continuously*, placing a tremendous stress on the driving SSR. The specter of this occurrence is possibly the most compelling reason to use a random-turn on SSR, especially when there is a high probability of load saturation, as in the case of regulating type transformer loads.

With a custom designed circuit, the turn-on phase angle can be tailored to suit the power factor of a specific load, but this is rarely done in practice. It would also be impractical in high volume production, unless it was made externally programmable; then expense would become a major consideration.

The zero crossing SSR is satisfactory for most applications, as evidenced by the large quantities in use today. Its low EMI qualities are very desirable, and it turns on in a predictable and consistent manner, thus quickly demonstrating its suitability for an application under the worst case load conditions. Since the turn-on point in a production SSR is not actually at zero voltage but more typically at 10 to 20 volts after zero, the previously mentioned hazards with inductive loads are not as real as they otherwise might be.

It is interesting to note that the zero switching circuit is functional only during the initial half cycle, and that both zero and random turn-on SSRs appear identical thereafter. While its continued presence can be a nuisance in some cases, the zero switching circuit provides a high level of noise immunity in the off state. It does this by clamping off the sensitive circuit areas after the zero window closes, during high line where noise is at its highest and the SSR is most vulnerable. This type of protection does not usually exist with other forms of triggering.

Some manufacturers advocate turn-on at the "peak" of the line for inductive loads, and such devices are available on the market (Fig. 8-7). For low power factor loads, these devices can reduce the surge current dramatically (compared to a zero switch) depending on the direction in which the load was last polarized, but they will not necessarily prevent the load from saturating. If coupled with a so-called "integral cycling" circuit, this approach would be more beneficial and far less likely to cause saturation. The peak firing SSR can be recommended in certain cases (power factors less than 0.5), but should not be used in general applications, especially for lamp and capacitive loads where nonlagging inrush currents may be enhanced and become destructive.

There are other switching techniques that may be employed for high inrush loads, but they are not usually found in production SSRs. The integral cycling circuit that delivers full line cycles in the same polarized sequence will ensure that the initial turn-on half cycle is of the opposite polarity to the last turn-off half cycle (Fig. 8-8). When SCR1 is triggered on during a positive half cycle, the energy stored in L1 will trigger on SCR2 early in the negative half cycle. This is excellent for transformer switching and will prevent load saturation while the circuit is intact. However, if on initial hookup the polarity of the first half cycle and that of a magnetized load happen to match, the load could saturate and instantly destroy the SSR.

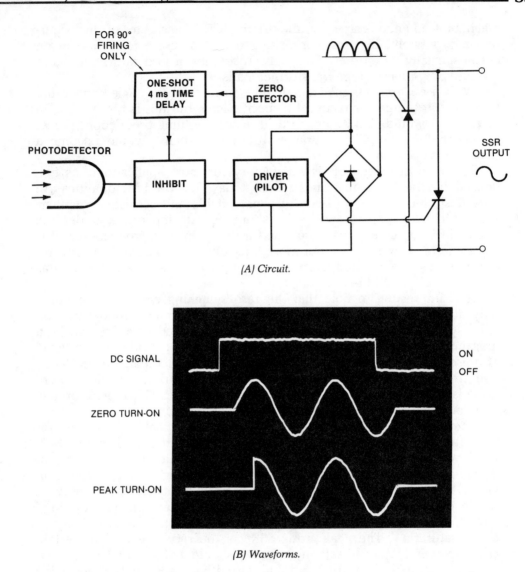

(A) Circuit.

(B) Waveforms.

Fig. 8-7. *Typical circuit configurations and waveforms for zero and 90° (peak) turn-on.*

In some versions of the integral cycle approach where a capacitor is used to store the trigger energy for the slave SCR, a third terminal is required in the output to access the generator side of the load to charge the capacitor, which may not be practical in some applications.

Another surge-reducing switching technique, and possibly the best, is the soft start shown with a typical waveform on an expanded scale in Fig. 8-9. With this system, once the control signal is applied, the SSR is ramped on by internal circuitry that advances the turn-on phase angle over several half cycles. The slow transition to full line voltage virtually

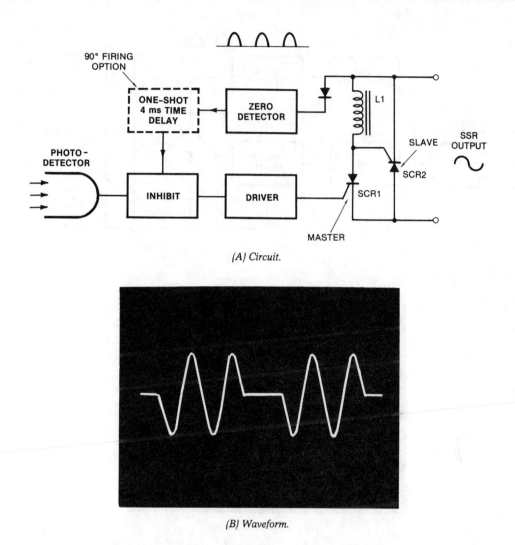

(A) Circuit.

(B) Waveform.

Fig. 8-8. *Typical circuit configuration and waveform illustrating integral cycle switching, with zero voltage turn-on.*

eliminates the problems associated with zero, random, peak, and integral cycle turn-on. It is also beneficial for lamps and capacitive loads and could be applied in most general applications.

While soft start spreads the inrush current over many cycles, thus reducing stress, it also prevents the occurrence of enormous saturating currents. Due to its phase control nature, it can produce a brief burst of EMI noise during the ramp-up period; possibly a small price for the benefits gained.

The soft-start approach has great appeal, but the prolonged turn-on period may not always be acceptable. Also, it is difficult to choose a ramped turn-on period that would be universally acceptable. This,

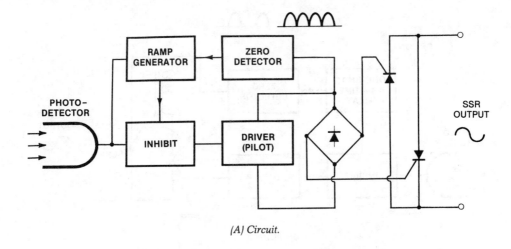

(A) Circuit.

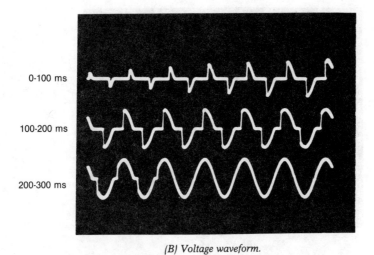

(B) Voltage waveform.

Fig. 8-9. *Typical circuit configuration and voltage waveform of soft start, with phase angle ramped on over 14 cycles.*

together with the higher cost, is probably the reason why SSRs of this type are not readily available. From the practical and economic standpoint at this time, the best choice may still be a standard SSR, oversized to withstand the high surges.

8.5 *Motor Switching*

Dynamic loads such as motors and solenoids, etc., can create special problems for SSRs, in addition to those discussed for passive inductors.

High initial surge current is drawn because their stationary impedance is usually very low. For example, after the initial surge, a solenoid core will pull in and "seal" at a much lower steady-state current, possibly by dropping to less than 25%. With motors, the change in current from stall to run can be even greater, possibly dropping to less than 20%, depending on the type.

As a motor rotor rotates, it develops a back emf that reduces the flow of current. This same back emf can also add to the applied line voltage and create "overvoltage" conditions during turn-off. Mechanical loads with a high starting torque or high inertia, such as fans and flywheels, will of course prolong the start-up surge period, which should be taken into account when selecting the driving SSR. When the mechanical load is unknown, as may be the case with a power tool, worst case conditions should apply.

Most of the surge-reducing techniques discussed in Section 8.4 can also be applied to motors. The soft-start method is particularly useful with universal and capacitive start types. Many small power tools and machines utilize a manually operated version of phase angle control to vary the motor speed. A similar technique is also used on motor speed regulation controls, where the phase angle is controlled by a signal fed back from a shaft speed detector.

Some motors incorporate an additional (start) winding that is centrifugally switched out when the rotor is up to speed, which can further exaggerate the start-up current (Fig. 8-10A). Centrifugal (mechanical) switches, especially governor speed controls, can also create problems in surrounding equipment, due to EMI noise generated by the arcing contacts. This type of switch would be best designed to operate the motor via the low control current of an SSR rather than switch the motor directly (Fig. 8-10B). When the control source for these functions is an electrical signal generally based on time, current, or motor speed, the SSR could completely replace the mechanical contacts. (Refer to Section 11.6.) Unless all connections are accessible, these options may only be available to the motor manufacturers.

A stalled motor continually passing the high locked-rotor current is always a possibility. This can be dealt with by means of a thermal overload switch or a fuse, requiring only that the driving SSR be designed to withstand this current for the device opening period.

It can be seen that motor loads are less predictable than passive inductive loads, with more possibilities to complicate the start-up condition. However, SSR selection can be simplified by using the tables shown, which provide a reasonably accurate rule-of-thumb method of SSR selection. The current values in Table 8-1 are typical and based on industry average worst case motor ratings, with efficiencies ranging from 20 percent for fractional types to 60 percent for power types. The value from the table is the run current of the motor and does not include the start-up surge. It should therefore only be applied to the motor rating of the SSR (if known). For ease of selection, Table 8-2 provides typical SSR motor

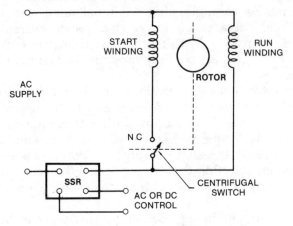

(A) Typical configuration of motor with start winding.

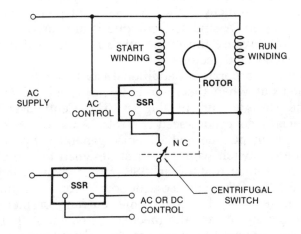

(B) Start winding switched via SSR to reduce EMI noise.

Fig. 8-10. *SSR control of motors.*

ratings versus the general use ratings. For greater accuracy, specific SSR motor ratings and actual motor run current should be requested from the manufacturers. Unfortunately, SSR manufacturers do not usually publish their motor ratings, which, as can be seen, are considerably reduced from the general purpose ratings. Most manufacturers have Underwriters Laboratories recognition on their SSRs; therefore, the motor ratings are based on the same UL test methods. However, it does not follow that all SSRs have the same ratings. The following table illustrates the differences in actual motor ratings between two well known manufacturers, with similarly specified parts.

The motor ratings in Table 8-1 are an average for the industry, intended only for general guidance in the absence of actual ratings. The

Published SSR Rating (General Use)	Manufacturer "A" (Motor Rating)	Manufacturer "B" (Motor Rating)
10 ampere	2.5 ampere	4.5 ampere
25 ampere	5.0 ampere	8.0 ampere
40 ampere	10 ampere	14 ampere

locked rotor value given in Table 8-2 is the true measure of the SSR surge capability, since this is the parameter that is tested and must comply with Underwriters Laboratory requirements. The general procedure for UL testing is that the SSR must survive a test current six (6) times the full load rating for one (1) second. The test is repeated fifty (50) times at a duty cycle of one (1) second on, nine (9) seconds off, with a 0.45 power factor load.

Table 8-1. *AC Horsepower Versus Full-Load Motor Run Current (Amperes)*

MOTOR	SINGLE-PHASE			THREE-PHASE			
H.P.	115 V	*230 V	440 V	115 V	*230 V	440 V	550 V
1/16	1.8						
1/10	2.5						
1/8	3.2						
1/6	4.0	2.0					
1/4	5.2	2.5					
1/3	6.5	3.2	1.8				
1/2	8.0	4.2	2.4	4.0	1.9	.96	.82
3/4	11.8	5.5	3.2	5.5	2.8	1.5	1.0
1	14	7.0	3.9	7.0	3.5	1.9	1.4
1-1/2	19	9.2	5.0	10.5	5.1	2.6	2.0
2	24	12.5	6.2	14	6.6	3.4	2.6
3	35	17	8.0	19	9.5	4.6	4.0
5	56	28	13	30	15	7.5	6.0
7-1/2	80	40	21	44	22	11	9.0
10		48	26	56	28	14	11
15		64	33		41	20	17
20			42			26	21
25			54				26

*208-volt rated motors 10% higher current.

While there is an AC solid-state relay for almost every application, the same cannot be said for DC SSRs. The DC SSRs available on the market today are very limited, the majority of which are rated at less than 4 amperes at approximately 60 volts. This being the case, very little statistical data exists from which to compute typical DC SSR motor ratings; therefore, these should be obtained from the SSR manufacturer. If motor ratings for the SSR are not available, the important parameters to consider in this case are the surge rating of the SSR, if any, versus the motor

Table 8-2. *Typical AC SSR Motor Ratings*

SSR GENERAL USE RATING AMPERES	MOTOR LOAD AMPS	
	Full Load (Run)	Locked Rotor (Start)
2	1.0	6.0
4	1.75	10.5
5	2.0	12
8	2.5	15
10	3.0	18
15	5.0	30
25	7.0	42
40	12	72
75	20	120

locked rotor rating. The surge rating for a DC type is usually far less than that of a similarly rated AC type.

Table 8-3 lists the typical DC motor steady-state run current versus horsepower, which like the AC motor ratings are based on industry average worst case conditions. Efficiencies range from 36 percent for

Table 8-3. *DC Horsepower Versus Full-Load Motor Run Current (Amperes)*

MOTOR HP	115 V	230 V	550 V
1/10	1.8	.9	—
1/8	2.0	1.0	—
1/6	2.3	1.1	—
1/4	2.9	1.4	—
1/3	3.8	1.8	—
1/2	5.0	2.5	1.2
3/4	7.0	3.5	1.5
1	9.0	4.5	1.9
1-1/2	12.8	6.4	2.6
2	16.6	8.3	3.5
3	24	12	5.0
5	40	20	8.2
7-1/2	60	30	12
10	78	39	17
15	112	56	25

fractional to 87 percent for power types. The conversion factor for horsepower to current, assuming 100 percent efficiency, is:

$$\frac{746 \times HP}{LINE\ E} = Amperes$$

As previously described, SSRs, when used in motor drive applications, require considerable derating. In applications involving motor braking or reversing, further derating or additional precautions may be necessary. Where a capacitor is involved, as in the case of a split-phase reversing motor (Section 11.7), a voltage-doubling effect may occur. This may also happen due to the back EMF of the motor itself when coasting. In either case, it would be wise to choose an SSR rated at twice the applied line voltage. Furthermore, if the capacitor circuit contains no limiting resistance in series with the SSRs, a resistance should be inserted. The values should be calculated to limit the possible surge current to at least the one cycle surge rating of the SSR.

In some circuits the possibility of a line to line short may exist due to (contact) overlap during transition, or a transient (dv/dt) turn-on at start up. If such is the case, this might be considered a hazardous or borderline application for an SSR, and therefore requires great care in design. The insertion of current-limiting resistors would be mandatory, with the possible need for a time delay-on-operate in each control circuit. The surge should again be limited to at least the surge rating of the SSR, while the delay to prevent overlap could be implemented in the drive logic.

It should be noted that overvoltage caused by capacitive voltage doubling or back EMF from the motor cannot be effectively dealt with by adding voltage-transient suppressors. Suppressors such as metal oxide varistors (MOVs) are typically designed for *brief* high voltage spikes and may be destroyed by sustained high energy conduction. It is therefore important that SSRs are chosen to withstand the highest expected sustained voltage excursion.

8.6 Lamp Switching

The inrush current characteristic of tungsten filament (incandescent) lamps is somewhat similar to the surge characteristic of the thyristors used in AC SSR outputs, making them a good match. The typical ten times steady-state ratings which apply to both parameters from a cold start allow many SSRs to switch lamps with current ratings close to their own steady-state ratings. Some lamps reportedly have even higher instantaneous inrush currents. This is rarely seen in practice since line and source impedances and filament inductance become significant at higher currents, all of which tend to limit the peak current. Generally, the ten times steady-state rating is considered a safe number for lamps.

Fig. 8-11 shows a typical low powered (less than 100 watt) lamp inrush characteristic compared to the allowable 1 cycle surge pulse for a typical SSR. It must be remembered that the SSR pulse is nonrepetitive and such a design would require further consideration if the lamp were flashed on and off at a rate too fast for the thyristor junction to cool off, say greater than 10 cpm. On the other hand, if the repetition rate became fast enough to prevent the lamp filament from cooling, the inrush current would be substantially reduced and thereby compensate for the

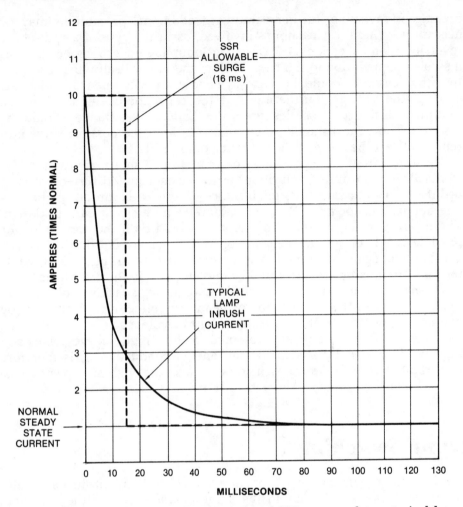

Fig. 8-11. *Current versus time characteristics of SSR compared to a typical lamp (less than one ampere) at turn-on.*

faster repetition rate. The cooling rate of the filament would depend heavily on the thermal inertia of the lamp, which increases with the lamp current rating.

At less than 1 ampere, lamps benefit from the life extending properties of the SSR zero voltage turn-on characteristics. The sinusoidal rise from zero voltage provides a soft start, compared to the thermal and mechanical shock experienced by the filament when turned on at peak line voltage. At higher currents, the thermal inertia of lamps causes an increasingly larger portion of the inrush current to occur over an increasing number of cycles, thus eventually the effect of zero voltage turn-on is nullified.

While most manufacturers' SSR lamp ratings are based on similar test criteria, as in the case of motor ratings, they vary somewhat for similarly rated parts. Some SSR lamp ratings are equal to the general purpose (resistive) ratings while others can be as low as 50 percent. Again, these values are not normally published and should be requested from the manufacturer. The Canadian Standards Association (CSA) prefers that the general purpose lamp and motor ratings are marked directly on the SSR, which is far more convenient and avoids confusion. U.S. manufacturers would be well advised to do the same. The generally accepted testing procedure for incandescent lamps, also prescribed by EIA/NARM and Underwriters Laboratories, is as follows: The SSR must survive a test current (resistive) that is 1.5 times the full load rating for one (1) second. The test is repeated 50 times at a duty cycle of one (1) second on, nine (9) seconds off.

SSRs are now being used more frequently for the phase control dimming of incandescent lamps, spurred on by the computerization of the stage lighting industry. A limited number of models are available with the control logic built in, which requires a variable resistance or voltage at the input to control the phase angle. With stage lighting systems, the timing logic is usually contained in the controller; therefore, the largest quantity of units sold are the "phase controllable" type. These are specially characterized nonzero SSRs, as opposed to standard random turn-on types which may not operate in the phase control mode. The difference is that the phase controllable types must respond to a pulse and operate over the widest possible phase angle range and recover every half cycle, whereas random turn-on types are only required to respond to a DC control signal and stay at full conduction. The waveform shown in Fig. 8-9B to illustrate "soft start" is representative of phase angle control, except that the controller in this case may adjust to any position and hold it indefinitely.

A newer series of lamps known as tungsten halogen or halogen cycle lamps are being used in applications where high intensity concentrated light is required. These lamps utilize a filament redepositing system that allows a much hotter than normal filament temperature, which in turn creates a much higher than normal inrush current. The surge can be up to 25 times steady-state current if conditions allow and should be taken into consideration when driving with an SSR. Furthermore, dimming this type of lamp should be avoided, since lowering the lamp temperature can defeat the filament redepositing system and seriously shorten the life of the lamp.

Other lighting systems are comprised mainly of the gas discharge variety (neon, fluorescent, mercury vapor, etc.), which use a variety of inductors, capacitors, and switching devices that affect the power factor. While the lamps themselves are mainly capacitive, the total load may appear inductive due to a series transformer or a ballast.

Some loads, such as metal halide lamps, are quite complex and go through wide power factor swings at turn-on. During lamp warm-up, a process that could take 10 minutes, the lamp and its ballast can appear inductive, capacitive, and finally resistive, accompanied by current spikes up to 100 times its rating. A combination such as this might require an SSR rated ten times higher than the lamp it is switching. Lamp loads of this type should be treated on an individual basis, with careful examination of manufacturers' data when available, and possibly a discussion with the SSR manufacturer.

9

Protective
Measures

Possibly the greatest shortcoming that SSRs have when compared to EMRs is their greater susceptibility to electrical noise, transient overvoltage, and overcurrent.

9.1 Noise Susceptibility

The noise susceptibility is the total responsibility of the manufacturer and is dependent on circuit design, sensitivity of components and stabilization techniques used. In the case of overvoltage and current, it is the responsibility of the user to ensure that the manufacturer's maximum ratings are not exceeded.

SSRs generally do not fail catastrophically due to noise, unless they happen to mistrigger during a point in the line cycle when an excessively high current surge (di/dt) might occur (e.g., a cold tungsten lamp filament can cause an instantaneous current flow greater than 20 times its rating at peak line voltage). Another rather remote possibility is that repetitive mistriggering of one polarity only might cause an inductive load to saturate and thereby draw currents destructive to both load and relay.

Usually, a malfunction due to noise is only temporary, such as turning on when the SSR should be off, and vice-versa. By its very nature, noise is difficult to define, being generated by the randomness of contact bounce and arcing motor commutators, etc. There are certain industry recognized tests that may be performed to determine a given level of susceptibility, although there appears to be very little standardization among manufacturers on this point. The following specifications contain the more commonly used susceptibility tests:

NEMA 1CS 2-230 Showering arc test.
EIA/NARM TS-443 Common mode noise test.

In both tests the noise generated by the tester is high frequency in nature—broadband for the former, and specifically 1.5 megahertz for the latter. This type of noise, more properly defined as electromagnetic interference (EMI), affects the SSR by feeding signals into the sensitive parts of the circuit, such as the pilot SCR or the base of the "receiver" transistor in the photocoupler. The noise is either conducted through the wiring or coupled through low value capacitances within the package; even the potting material can be a contributor. PC board layout and spacings will strongly influence susceptibility to this type of noise.

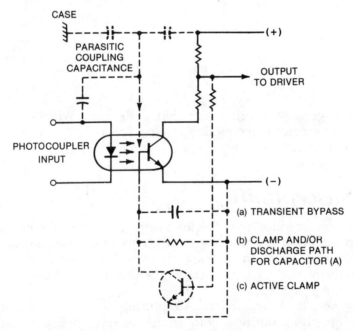

Fig. 9-1. *Corrective measures for RF parasitic coupling into base of phototransistor.*

Fig. 9-1 shows how some manufacturers deal with parasitic noise coupling into the base of a phototransistor. The RC combination of (a) and (b) are most commonly used, while the active clamp (c) improves high line stability but still requires the capacitor to nullify injected noise when unclamped (close to zero crossing). The commonly built-in snubber RC network across the output is also effective in reducing sensitivity to noise, especially at lower frequencies. These solutions are similar for those circuits using photo-SCRs and pilot SCRs, together with special biasing techniques to stabilize the SCR gain.

Some of the techniques used to reduce noise in the coupler and drive circuits are also effective against false triggering caused by voltage transients on the input lines. When a capacitor is added, for example, the response time which is not critical for AC SSRs may be lengthened, possibily from a few microseconds to tenths of milliseconds. Because of this delay, voltage transients or bursts of shorter duration are rejected, thus improving noise immunity in respect to time. A further limitation on input transient susceptibility is imposed by the narrowness of the zero switching window, which, once closed, will reject such false signals for the balance of the half cycle. Some specialized SSRs, such as those designed for traffic control applications, have high input on-off thresholds to defeat control line noise (e.g., turn-on at 12 volts, turn-off at 8 volts).

Most AC SSRs use thyristors in their drive and output circuits which, due to their regenerative nature, can latch on for a whole half cycle when triggered by a brief high-voltage transient, thus acting as a pulse stretcher. In addition to responding to the amplitude of the transient, a thyristor can also mistrigger when the rate of rise (dv/dt) of a transient or applied voltage exceeds certain limits, as discussed in Sections 3.2 and 9.2. Transient suppressors are effective against the former, and the RC snubber improves the tolerance of an SSR to the latter. The dv/dt problem would be nonexistent if the output device was a nonlatching bidirectional switch, such as a bipolar transistor used in a full-wave bridge (Section 11.4). However, cost, surge and dissipation would be serious limiting factors in this case.

9.2 dv/dt (Rate Effect)

The rate effect phenomenon in thyristors (dv/dt) is caused by capacitive coupling within the device structure between the high terminal (anode) and the gate, which could result in self-induced turn-on if the SSR dv/dt limits are exceeded. The expression dv/dt defines a rising voltage versus time expressed in volts per microsecond (V/μs). When applied to an AC SSR as "static" or "off-state" dv/dt, it is a parameter that defines the minimum dv/dt withstand capability of the SSR, or in other words, the maximum allowable rate of rise of voltage across the output terminals that will not turn on the SSR (typically 200 V/μs).

The static dv/dt measurement is made by applying a linear voltage ramp up to a given voltage (V_p), or an exponential waveform asymptotic to the applied voltage (V_p) as shown in Figs. 9-2A and B. Either method is acceptable, the latter being the easiest to implement with a simple RC network and therefore the most common. The period from zero time to one time constant (TC) is considered to be the linear portion of the exponential ramp. The ramp in each case indicates the rate of change (dv/dt) below which the SSR will not be mistriggered (shaded area).

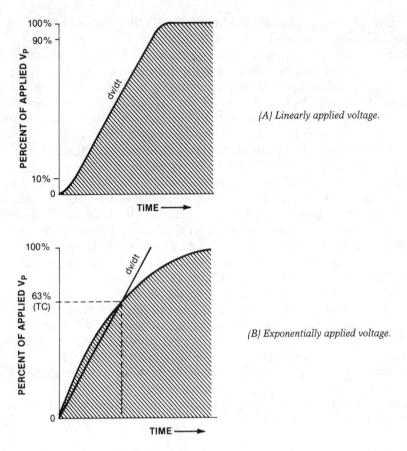

(A) Linearly applied voltage.

(B) Exponentially applied voltage.

Fig. 9-2. *Test waveforms for dv/dt measurements.*

Some caution is advised when checking dv/dt specifications, in that the statement "at rated voltage" may actually mean transient peak, steady-state peak, or 63 percent of the steady-state peak value (e.g., a test using Fig. 9-2A would be more severe than a test using 9-2B, if the same applied voltage (V_p) and ramp values were used). The actual target voltage of the ramp is important because SSR dv/dt withstand capability decreases as voltage amplitude increases.

In addition to the foregoing, there is often confusion about the dv/dt testing of SSRs. For example, if a given voltage ramp is applied directly across the SSR output without regard to circuit impedance, the effectiveness of the snubber will be zero because the source impedance is assumed to be zero. In practice, of course, the load, line and power source all have impedance that the snubber can work across; therefore, a load-source impedance should also be stated for a dv/dt value to be meaningful. Most manufacturers use the EIA/NARM Standard RS-443 value of 50 ohms for consistency in testing and in published data (Section 12.10).

The commutating dv/dt parameter found in triac specifications refers to the triac withstand capability to the rate of rise of reapplied voltage immediately after conduction, as shown in Fig. 9-3A (i.e., the device's ability to regain blocking), typically in the order of 5 to 10 V/μs. The phenomenon occurs with inductive loads where the current is lagging the voltage (Fig. 9-3). When the triac turns off at zero current, the voltage, which has advanced into the next half cycle, instantly appears across the triac. It is this rate of rise of voltage that must be limited below the stated value or retriggering may occur, causing the triac to lock on.

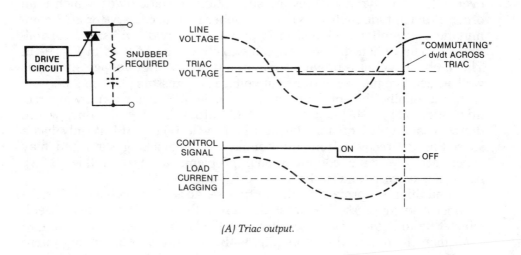

(A) Triac output.

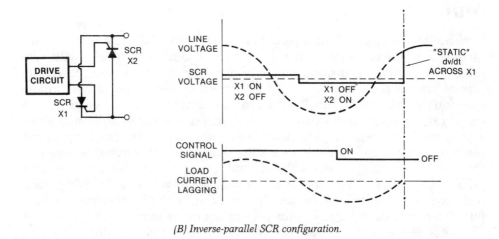

(B) Inverse-parallel SCR configuration.

Fig. 9-3. *Turn-off conditions for triac and SCRs switching inductive loads.*

Where the critical operating limits of an SSR are specified, including load power factor, temperature range, etc., the effect of commutating dv/dt has already been considered in the design and is of no concern to the user, and therefore it is not specified as a parameter. When the output device is a triac, a snubber is considered essential for inductive loads and will almost certainly have been included so that performance will equal that of dual (inverse-parallel) SCRs.

Dual SCRs do not have the commutating dv/dt problem that triacs have since each device has a full half cycle to turn off (Fig. 9-3B). However, like most thyristors, they are susceptible to static dv/dt, which is an order of magnitude higher. While a snubber is unnecessary for SCR commutation, it is often included to improve static dv/dt withstand capability, especially when driven by a gate-sensitive pilot SCR as in Fig. 9-4B. The snubber will also provide some immunity to brief voltage spikes, as well as facilitate the use of transient voltage suppressors (Section 9.3).

Occasionally, the configuration of a single SCR inside a full-wave rectifier bridge (Fig. 9-4A) is used as an SSR output. Unlike pilot duty where the SCR is shunted off for almost a half cycle (Fig. 9-4B), it only has a short time to recover from the conducting to blocking state and may experience the same commutational problems as a triac (Fig. 9-3A). Again, a good case for a snubber.

In addition to protecting the output devices from voltage spikes, a snubber also suppresses dv/dt (static) across the pilot SCR (where used), which due to higher trigger sensitivity is likely to be more susceptible to dv/dt than the output thyristors (Fig. 9-4B). Thus, a substantial argument exists for the inclusion of a snubber regardless of the output thyristor type or configuration used.

9.3 Snubbers

The internal RC network (snubber) used in AC SSRs is a major factor in transient voltage and dv/dt suppression. It deals effectively with two facets of a voltage transient; not only does the network slow down the rate of rise as seen by the output thyristors and sensitive drive circuits (previously discussed), but it also limits the amplitude to which it can rise. In the latter case, however, the protection is somewhat limited since a prolonged transient or pulse train will eventually "staircase" up to the blocking voltage and possibly cause breakdown. In this event, a suppressor with a specific clamp voltage, such as a zener diode or varistor (MOV), is called for. The snubber will still serve well as a clamp by holding down the leading edge of the transient during the brief period it takes for the suppressor to turn on, thus preventing overshoot.

While the snubber properties are mostly beneficial, it does give rise to a substantial AC component in the off-state leakage. For a typical AC SSR with say 5 milliamperes of off-state leakage, possibly half, or 2.5 milliamperes may be contributed by the snubber. The AC portion of the

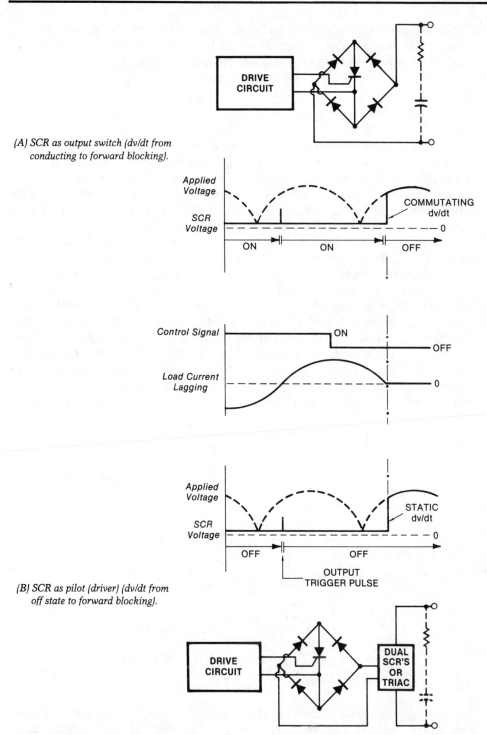

(A) SCR as output switch (dv/dt from conducting to forward blocking).

(B) SCR as pilot (driver) (dv/dt from off state to forward blocking).

Fig. 9-4. *Turn-off conditions for SCRs in full-wave bridge circuits switching inductive loads.*

leakage current is proportionately related to frequency; therefore, any attempt to use an SSR designed for 60 hertz at a higher frequency, say 400 hertz, would increase the leakage current accordingly:

$$\frac{400 \times 2.5}{60} = 16.67 \text{ milliamperes}$$

It would be unwise to operate any device beyond the manufacturer's specified limits and expect it to work. However, when prototyping with available parts, it might be worthwhile knowing that a 60-hertz characterized SSR with inverse-parallel SCRs in the output will most likely substitute for a 400-hertz part. On the other hand, an SSR with a triac output can get into commutational difficulties and lock on.

Snubber design for limiting both the commutating and static dv/dt associated with thyristors can be complex and time consuming. The circuit elements involved are an RC network across the output thyristor, working in conjunction with the resistive and reactive characteristics of the load. The circuit illustrated in Fig. 9-5 is a damped tuned circuit which may receive a voltage step at initial application of power or when the thyristor turns off, the amplitude of the latter depending on the load power factor. Capacitor C1 can increase the SSR dv/dt withstand capability to this fast rising voltage, while resistor R1 can increase damping and reduce resonant overshoot. With inductor L1 in the load, the softening effect of its transient impedance will allow a larger value of R1 and a smaller value of C1.

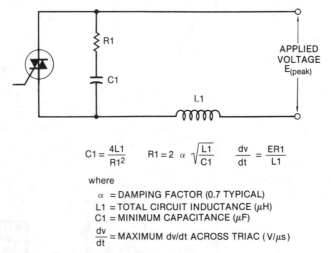

$$C1 = \frac{4L1}{R1^2} \qquad R1 = 2 \; \alpha \; \sqrt{\frac{L1}{C1}} \qquad \frac{dv}{dt} = \frac{ER1}{L1}$$

where

α = DAMPING FACTOR (0.7 TYPICAL)
L1 = TOTAL CIRCUIT INDUCTANCE (μH)
C1 = MINIMUM CAPACITANCE (μF)
$\frac{dv}{dt}$ = MAXIMUM dv/dt ACROSS TRIAC (V/μs)

Fig. 9-5. *SSR snubber network.*

For a given load situation, the values in Fig. 9-5 can be idealized; however, such is not the case with a commercial SSR which must perform equally well under a wide variety of load conditions. The manufacturers of packaged SSRs take care of this design chore, generally including snubber capacitor C1 that is as large as physically possible to enhance the SSR dv/dt rating and transient absorption properties. Resistor R1 is chosen to limit to a safe value the initial capacitor discharge current through the driving pilot SCR and output device during the turn-on interval.

The SSR user, with no control over snubber values, has limited options should an unfortunate combination of snubber and load result in ringing or resonant overshoot. Where the SSR peak blocking (transient) ratings are threatened or exceeded, the options are to add to the snubber externally, use a higher voltage rated SSR (however, snubber capacitor value may be lower), or use a transient voltage suppressor. In many cases the SSR may safely self trigger (anode fire) thus protecting the capacitor and voltage sensitive components from damage.

The specified parameters that affect the SSR's usage in regard to snubber values are the dv/dt (static) rating and the power factor (inductive load) rating (typically 200 volts per microsecond for the former and 0.5 for the latter). Typical commercial SSR snubber values are shown in Table 9-1.

As a point of interest and also a reason for keeping the snubber resistor small, a fast rising voltage step applied to the network of Fig. 9-6 will cause a similarly fast rising voltage to appear across the triac. The value to which this voltage (E2) rises before the snubber becomes effective (C1 discharged) is determined by the source voltage (E1) and the ratio of R1 to the total series impedance (R1 + R2 + R3). After the initial step, which hopefully does not mistrigger the triac, the voltage rises logarithmically towards E1, defined by the time constant of C1 and the total series resistance. Generally, the initial rise is small and probably below the triac dv/dt threshold (i.e., the voltage above which the triac becomes susceptible to dv/dt).

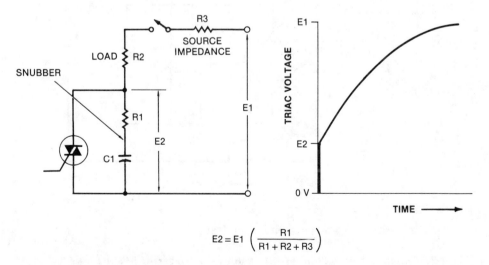

$$E2 = E1 \left(\frac{R1}{R1 + R2 + R3} \right)$$

Fig. 9-6. *Instantaneous voltage rise across triac before snubber becomes effective.*

Table 9-1. *Typical SSR Snubber Values*

AC RATING (V_{RMS})	CURRENT RATINGS (AMPS)	RESISTANCE (OHMS)	CAPACITANCE (μF)
120	10–40	33	.068
240	10–40	33	.033
480	10–40	47	.022

If the load is inductive, the voltage step and its rate of rise are limited by the high initial reactive impedance of the load. In the dynamic (running) mode where a voltage step is produced at triac turn-off due to lagging current, the rise is also influenced by the transient impedance of the inductor, the same inductor which ironically caused the step in the first place.

While the typical internal snubber value and the typical dv/dt specification are adequate for most applications, they may not prevent what is commonly referred to as the "blip" or "bleep" problem which occurs during start up. That is, when power is initially applied to the SSR/load combination usually by means of a mechanical switch, the resultant fast rising transient may mistrigger the SSR and possibly "let through" a half cycle pulse. Fortunately, most loads are not troubled by this pulse; however, if it should become a problem and it is not possible to increase the size of the snubber or the response time of the load, the circuit shown in Fig. 9-7 may provide a solution for lighter loads such as fast acting solenoids and counters, etc. This technique requires the insertion of some resistance (R1) in series with the SSR and the load, approximately equal to one tenth of the load resistance for the clamp circuit to "work across." Capacitor C1 in the full-wave bridge is initially discharged by R2 and

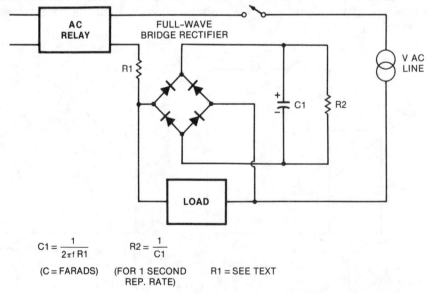

$$C1 = \frac{1}{2\pi f \, R1}$$

$$R2 = \frac{1}{C1}$$

(C = FARADS)   (FOR 1 SECOND REP. RATE)   R1 = SEE TEXT

Fig. 9-7. *Clamp circuit to eliminate thyristor false triggering at start up.*

absorbs the first half cycle pulse by charging up. The time constant of C1 and R1 is adjusted to reduce this pulse below the threshold of the responding load. An additional feature of this circuit is that it will track the peak line voltage and offer a low impedance to any subsequent rapidly rising voltage or would-be transient.

9.4 Suppressors

For those situations where overvoltage transients occur, possibly mistriggering the SSR, some form of suppression beyond the capabilities of the snubber is called for. The usual technique is to add a clamping device across the SSR terminals that will absorb the transient energy above a predetermined level. Over the years, the need for such devices that provide sustained voltage clipping has been filled by spark gaps, selenium and gas-discharge devices, together with the more common zener diodes and metal oxide varistors (MOVs).

When activated, some suppression devices, such as spark gaps, will drop to a very low impedance, thus transferring most of the transient to the load as if the SSR had turned on. Others, such as zeners and MOVs, will conduct only at the predetermined level and above, thereby sharing the transient with the load, which is the more common approach. If in a particular application it is deemed unacceptable for the load to receive any transient energy, the only solutions may be suppression of the transient source, or an SSR with a blocking capability higher than the transient.

Fig. 9-8 illustrates typical methods of suppressing transients across the SSR output "contacts," as well as suppression of transients at the source, which can be the load itself for DC inductive type loads. For the lower powered DC SSRs (less than 3 amperes, 60 volts), a single one-watt zener diode (commonly built-in) can provide adequate protection. Its value (V_Z) must be greater than the operating but less than the breakdown voltage of the SSR. For higher rated DC SSRs, zeners become impractical due to dissipation so more attention should be paid to suppression at the load, where the choice of a so-called "arc-suppression" diode may be simpler.

9.5 Diodes and Zeners

The diode shown across the load in A of Fig. 9-8 is the most effective way of suppressing the possibly hundreds of volts of back EMF that can be generated by the coil at turn-off. The disadvantages of this method are the SSR is not protected from other transient sources, and the drop-out time of the load (solenoid, clutch, EMR, etc.) may be extended by several milliseconds. If a prolonged drop-out time is objectionable, it may be considerably reduced by the diode-zener circuit in B of Fig. 9-8. The diode in each case should have a voltage rating (PIV) greater than the

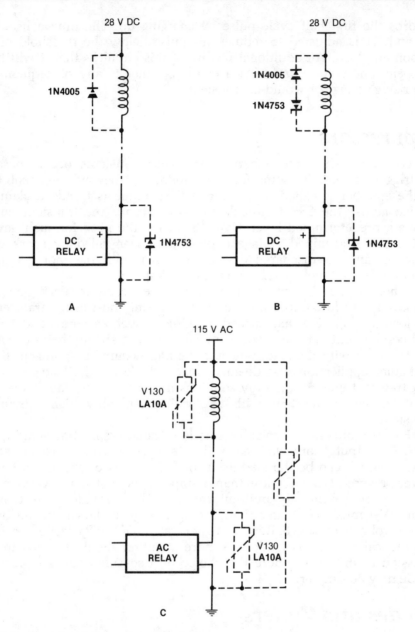

Fig. 9-8. *Transient suppression techniques.*

maximum operating load voltage, while the zener value (V_Z) should be no greater than the SSR breakdown voltage minus the load voltage. Some allowances should be made for the higher zener values that occur at peak (pulse) currents and at elevated temperatures.

The general rule in the selection of protective diodes and zeners is that their peak nonrepetitive (pulse) current ratings (Fig. 9-9A) should be equal to or greater than the maximum load current. Conservative steady-state power ratings for these devices may be ascertained from the following equation:

$$P_{watts} = \frac{I_L^2 L}{t_r}$$

where
  I_L = load current in DC amperes
  L = load inductance in henrys
  t_r = on/off repetition rate in seconds

Example: A load with a resistance of 4 ohms and an inductance of 0.0025 henry is driven from a 28-volt DC supply while being switched on and off 5 times a second:

$$I_L = \frac{28 \text{ volts}}{4 \text{ ohms}}$$

$$= 7 \text{ amperes}$$

$$t_r = \frac{1}{5} = 0.2 \text{ second}$$

$$P = \frac{7^2 \times .0025}{0.2}$$

$$= 0.613 \text{ watt}$$

A protective diode or zener with a 3/4-watt rating would suffice.

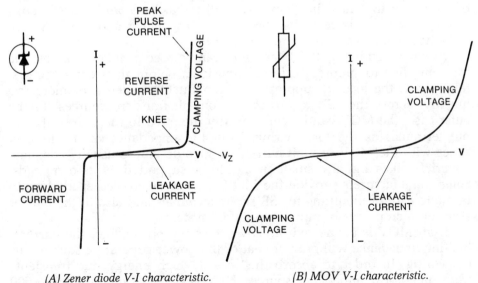

(A) Zener diode V-I characteristic.   *(B) MOV V-I characteristic.*

Fig. 9-9. *Comparison of zener diode and MOV characteristics.*

The zener diode is the ideal choice for protecting low voltage DC SSRs (less than 100 volts DC) used in parallel with the output. The zener "knee" is sharp and at lower voltages its clamping voltage changes little with current, a relationship that deteriorates proportionately as zener voltage increases. In the forward current mode (reverse for the SSR), the zener diode typically clamps as a single diode would at approximately one volt, thereby providing added reverse-voltage protection. When two zeners are used back-to-back (in series) with equal standoff voltages, they can be used to protect SSR outputs bidirectionally when switching AC loads. Some manufacturers provide the two zeners in a single package, characterized specifically for transient voltage protection. At higher voltages (greater than 100 volts) AC or DC, economics versus performance may suggest another of the many transient protective devices on the market today, the MOV (metal oxide varistor) being the most popular.

9.6 MOVs

The metal oxide varistor was developed about the same time as the SSR and has subsequently become a trustworthy companion of the SSR, providing much needed protection in some of its more hostile environments. MOVs, which are relatively new compared to zener diodes, are sold under a variety of names usually including the initials MOV, such as Movistor, Z-MOV, etc. As the name implies, its composition is that of metal oxides, primarily zinc oxide, sintered (fused) by high pressure and temperature into a ceramiclike material. The medium power types, ranging from 3 to 300 joules (watt/seconds), are generally formed in the shape of a disc 0.3 to 1 inch in diameter. With radial leads and a hard epoxy coating, they closely resemble disc ceramic capacitors and can be physically mounted in a similar manner (Fig. 9-10).

As suggested in Fig. 9-8C, the MOV can be used as follows: across the incoming line to suppress external transients before they enter the system; across the load to suppress load generated transients; or more frequently, across the SSR to protect it from all transient sources. In the latter case, the MOV can be conveniently mounted to the same SSR output terminals as the load wiring. With the impedance of the load in series with the MOV to limit current, a 30-joule unit is usually adequate for brief spikes and also small enough to be supported by its own leads. Some manufacturers provide the MOV with preassembled spade lugs to facilitate ease of mounting to SSR screw terminals and also to the barrier strips of microprocessor input-output (I/O) systems.

If an MOV is connected directly across the power line, the current-limiting impedance will only be that of the power generating source plus the wiring. In order to absorb the possibly high energy line transients from such a low impedance source, the larger chassis mount (300 to 600 joule) variety of MOV may be required. The greater expense of such a

Fig. 9-10. _Broad range of packages for metal oxide varistors (MOVs) are shown together with radial type (inset) commonly used with SSRs (70 joule, 20 mm unit). (Courtesy General Electric Company)_

device might be justified in that suppression across the line is required in one place only. Spark gaps might also be considered, and may even be a better solution in this heavy duty application.

Load generated voltage transients are generally not a problem with the sinusoidal waveform of the AC line and the typical switching characteristics of the SSR. However, an MOV can be used effectively across such loads as transformers and switching power supplies where spikes, too fast to be absorbed by the transformer itself, may be fed back into the primary (SSR load) winding. Used across the load, the MOV will not protect the SSR against line-borne transients; in fact, it may enhance them by shunting the filtering effect of an inductive load. While a particular situation might dictate where an MOV should be used, the first choice appears to be across the SSR terminals.

The electrical characteristics of the MOV are essentially that of a voltage-dependent, nonlinear resistor. When properly applied, the MOV impedance at nominal steady-state voltages should be so high that only a few microamperes of "leakage" current will flow. As a voltage of either polarity approaches its clamping voltage, the MOV rapidly becomes a low impedance, thus causing current flow to increase at a much faster rate than the increase in voltage. The amplitude of the instantaneous transient voltage across the SSR is the ratio of the IR drops across the MOV and the load-source impedances.

The voltage-current characteristics of the MOV and a zener diode are shown together in Fig. 9-9. The MOV characteristics could also be representative of back-to-back zeners. While the "knee" region of the MOV is less sharp than that of the zener, it does have the advantage of lower cost and higher voltage capability (up to 1200 volts).

To select the correct MOV for an application, it is desirable to have some knowledge of the source impedance and the pulse power of the transient. In the case of internally generated system transients, some fairly close estimates can be made when circuit parameters are known, and empirical data and waveforms are available.

For incoming line-borne transients, however, MOV selection is far more difficult since the nature of the transients and the characteristics of the power source are generally unknown. With very little data to go on, assumptions are often made using tables developed by various agencies on the amplitude and rate of occurrence of typical power-line voltage transients in various locations over long periods of time (Fig. 9-11). Assumptions on waveshape are also made by using industry standard pulses, typically 20 to 50 microseconds long, taken from existing published data such as IEEE Standard 28. A suggested source impedance is sometimes included, usually in the region of 50 to 150 ohms; alternatively, an impedance may be estimated from the KVA rating of the power supply (if known). In general, MOV selection for the power-line transients is often little more than an educated guess.

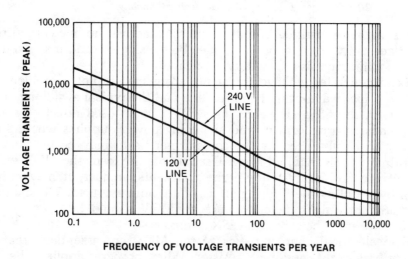

FREQUENCY OF VOLTAGE TRANSIENTS PER YEAR

Fig. 9-11. *Typical frequency of occurrence versus transient voltage amplitude for 120-volt AC and 240-volt AC power lines.*

When the MOV is placed across SSR output terminals, it has the increased security of a minimum known load impedance, in addition to the possibly unknown source impedance is series with it. Any inductance in the load will tend to further increase its impedance to fast rising transients and enhance the effectiveness of the MOV. As previously mentioned, a 20 to 30 joule MOV placed across the output terminals will be satisfactory for most applications, according to the collective field experience of a number of SSR manufacturers.

In selecting the MOV for voltage, its maximum continuous RMS voltage rating should be just above that of the highest expected line excursion, say 130 volts RMS, for the nominal 115-volt line, and 250 volts RMS for the nominal 220-volt line. To ensure complete protection the SSR transient (blocking) voltage rating must be greater than the MOV clamping voltage; that is, in order to divert the transient, the MOV must be conducting heavily in the ampere region before reaching the SSR breakdown voltage. (See the typical V-I curves in Fig. 9-12).

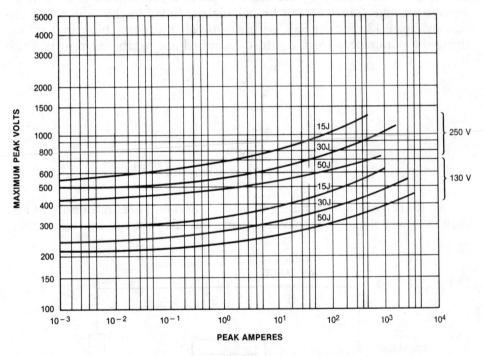

Fig. 9-12. *V-I characteristics of typical MOVs for use with SSRs.*

From the V-I curves, it can be seen that in order to accommodate the MOV, the SSR transient (blocking) voltage rating must be higher than is normally required for line operation (e.g., 400 volts for the 115-volt line and 600 volts for the 220-volt line). Most SSR manufacturers provide these higher voltage models (if required) for use with MOVs, but would normally suggest using a 220-volt model for the 115-volt line.

Some manufacturers recommend or even supply an MOV for use with a specific SSR, and also reference a test specification with a well defined transient that the combination is claimed to meet. In the absence of specific industry standards, where MOVs are specified for unknown conditions, it is good practice to reference an existing transient test specification where duration, waveform, and source impedance are accurately defined. Such a measure ensures that the values chosen will perform reliably in a given situation, and also allows reproducibility by providing a standard for comparison.

A commonly used test is the IEEE Standard 472 (1974) Surge Withstand Capability (SWC) test. The transient described in this test is an oscillatory (ringing) burst starting at 3 kilovolt peak and decaying to 50 percent in 6 microseconds at a frequency of 1.0 to 1.5 megahertz. The transient is applied to the SSR for a minimum of 100 bursts for 2 seconds from a generator with a source impedance of 150 ohms.

When the source impedance is known, the minimum peak current that must be conducted by an MOV to prevent breakdown of an SSR for a given voltage transient can be calculated. Refer to the equivalent circuit in Fig. 9-13. For an SSR with a 400-volt transient (blocking) rating, using the parameters from the test described above, the current is given by:

$$I_{MOV} = \frac{E_S - E_{SSR}}{Z_S}$$

$$= \frac{3000 - 400}{150}$$

$$= 17.33 \text{ amperes min.}$$

An MOV with appropriate values (i.e., 17.3 amperes minimum at less than 400 volts) can be selected from the manufacturers' V-I curves.

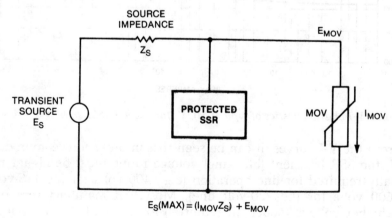

Fig. 9-13. *MOV/SSR equivalent circuit.*

Alternatively, the load line of the transient voltage and source imped-ance can be plotted directly in the logarithmic scale of the V-I graph, which will intersect with the proper MOV curve at a voltage below the transient rating of the protected SSR.

In an actual application, the SSR has the impedance of a load, the source impedance, and possibly other circuit impedances (e.g., wiring, etc.) in series with it, the sum of which should be considered in MOV selection.

Lifetime of the MOV is not infinite and manufacturers' pulse lifetime ratings should be consulted, especially where transients are known to exist and are also predictable. The curves shown in Fig. 9-14 indicate how an MOV of a particular size is derated according to the number of transient pulses anticipated during the equipment lifetime (Fig. 9-11). These curves are also useful for derating MOV peak current for single pulses of longer duration than the standard 8 X 20 microsecond pulse that this rating is based on (i.e., 8 microsecond rise and 20 microsecond fall to 50 percent from ANSI Std C62.1).

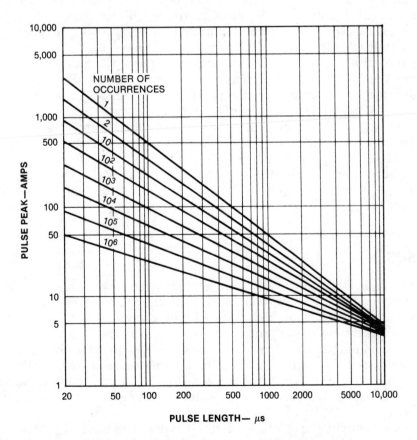

Fig. 9-14. *Typical lifetime curves of peak voltage versus duration (30 joules). End of life defined as change of ± 10 percent in clamping voltage.*

In addition to the allowable number of pulses during the MOV lifetime, the repetition rate may also become a factor if the transients occur in rapid succession. A pulse train can cause the energy requirements of the MOV to increase radically, by as much as the product of the single pulse energy (watt/seconds) and the number of pulses per second. The MOV selected must have a continuous (average) power dissipation rating (P_{AV}) equal to, or greater than, that created by the pulse train (e.g., a P_{AV} rating of 0.4 watt would allow only .006 joule per pulse at a repetition rate of 60 hertz).

It should be noted that the energy rating in joules is based on a square pulse of 10 milliseconds duration, as opposed to the shorter 8×10 microsecond pulse used for other parameters. In some data sheets, this pulse is allowed 1000 times in the life of an MOV, for a maximum change of ± 10 percent in clamping voltage.

Individual MOV specifications should be consulted for precise information regarding energy absorption, clamping properties, and physical size, since the relationships of these parameters will vary from one manufacturer to the next. When the MOV is used beyond its specified ratings, it will eventually fail, usually in the shorted mode. Failure, of course, is a positive indication that a higher energy device is called for. If failed shorted, the MOV will in its final act have preserved the SSR it was protecting, and is one of the arguments against "building it in." In failing, the MOV might also shatter and in so doing, possibly arc and carbonize its surroundings, which is possibly an argument in favor of encasing the MOV. Violent failure of this type, however, does not usually occur in the MOV or the SSR except in the unlikely event of a simultaneous transient and breakdown in the load. Appropriate fusing for the SSR or the MOV itself may be employed where the rupture of either one could be hazardous. Used within its ratings, the MOV will most likely outlive its associated equipment and provide low cost protective insurance for the SSR.

9.7 Fuses

Mechanical circuit breakers are generally too slow to prevent damage to semiconductor devices, and barely fast enough to protect printed circuitry and wiring from high short-circuit currents. Standard cartridge fuses are also too slow to protect semiconductors when operating close to their maximum ratings. However, they can be successfully used if the semiconductor is severely derated to accommodate a lower value fuse and "let through" current that the semiconductor can tolerate. Standard fuses are also extensively used to protect wiring, PC boards, and overall systems.

Semiconductor fuses are specialized types, designed to protect semiconductors while operating at close to their full ratings (Fig. 9-15). They are also referred to as current-limiting fuses, providing extremely fast opening, while restricting let-through current far below the available fault current that could destroy the semiconductor, as illustrated in Fig. 9-16. This type of fuse tends to be large and expensive, but it does provide a means of fully protecting SSRs against high current overloads where survival of the SSR is of prime importance. The fuses feature noncharring and nonarcing ceramic bodies for increased reliability.

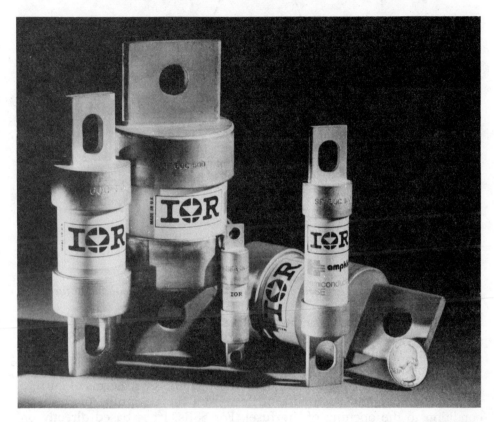

Fig. 9-15. *Typical semiconductor protective fuses from 5 to 800 amperes RMS. (Courtesy International Rectifier)*

The power system short-circuit capacity will influence how long it will take the protective fuse to interrupt the fault current, which is usually in the subcycle region for semiconductor types. Fuse opening time decreases with higher available short-circuit currents.

Referring again to Fig. 9-16, it can be seen that the fuse opening time has two modes, the initial "melting time" while current is rising, and the "arcing time" where current continues to rise to peak before decaying to zero. The sum total is referred to as the fuse "clearing time." The "peak available current" is that which would flow had the fuse remained closed.

In the protection of SSRs it is essential that the let-through current of the fuse during fault conditions remain within the surge withstand capa-

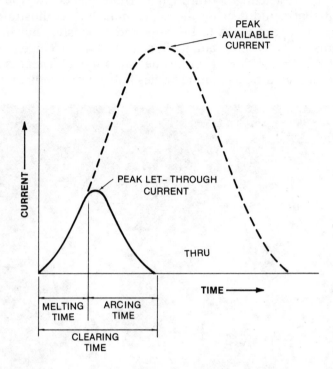

Fig. 9-16. *Illustration of subcycle current-limiting action of semiconductor fuse.*

bility of the output semiconductor (usually a thyristor for AC SSRs). In order to coordinate the fuse ratings with those of the SSR, the SSR data sheets generally use similar terms for fusing, such as I^2t (A^2s), a common expression in defining fuse performance.

For fuses, I^2t is the measure of let-through energy in terms of current versus time (i.e., the RMS current flow from the beginning of the fault condition to the opening of the fuse). For SSRs, I^2t is based directly on the output thyristor's single-cycle peak surge current determined by:

$$I^2t = \left(\frac{I^2_{PK(SURGE)}}{2} \right).0083_{(SECONDS)}$$

The procedure is to select a fuse with an I^2t let-through rating that is less than the I^2t capability of the SSR, for the same duration.

As a matter of interest, some manufacturers suggest that in projecting SSR I^2t ratings into the subcycle region to match the fuse let-through ratings, the formula $I^2\sqrt{t}$ would be more appropriate than the traditional I^2t as a constant. The $I^2\sqrt{t}$ rating of the SSR can be used to determine the I^2t rating of the SSR for whatever time it takes the fuse to clear:

$$I^2t \text{ at time tx} = I^2\sqrt{t} \times \sqrt{t_x}$$

The rationale for $I^2\sqrt{t}$ is that these values more closely follow the thyristor transient thermal impedance curve, especially in the subcycle region. The $I^2 t$ and $I^2\sqrt{t}$ values for a typical 25 ampere SSR, when plotted against its maximum surge current (in peak values) in Fig. 9-17, seem to confirm this.

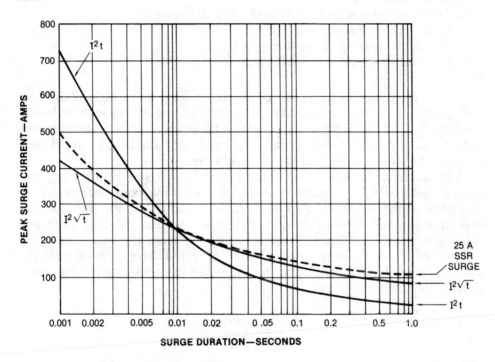

Fig. 9-17. *$I^2 t$ and $I^2\sqrt{t}$ compared to typical 25-ampere SSR surge characteristic (based on 8.3 millisecond value).*

As a further point of interest, the SSR surge ratings are usually based on thyristor specifications that stipulate an after-surge reverse-blocking capability, even though most SSR specifications do not reflect this. When a series fuse opens, there is no further demand for conduction or blocking and consequently the thyristor can withstand a higher surge than the published ratings. The surge current for a thyristor is approximately 20 percent higher if no voltage is applied immediately after a current surge lasting one half cycle or less. This means that SSRs can be considered to have higher current capability and therefore higher $I^2 t$ (or $I^2\sqrt{t}$) ratings when protected by a semiconductor fuse. The 20 percent higher values may be considered as built-in conservatism when selecting the right fuse.

Another factor that affects the fuse to SSR match is that their ratings are based on different waveshapes, triangular for the former and sinusoidal for the latter. However, the difference is small and in any case the thermal effect of the fuse let-through $I^2 t$ will be less than that of an equal $I^2 t$ value for the SSR; this again is on the side of conservatism.

The following are the main parameters used in the selection of a semiconductor fuse:

1. Fuse voltage rating
2. Fuse current rating
3. Available system fault current (Source E/Source Z)
4. Fuse peak let-through current, I_{plt}
5. Fuse total clearing (or let-through) I^2t
6. I^2t or surge withstand capability of SSR.

The fuse voltage rating must, of course, at least be as high as the system voltage, and the current rating greater than the steady-state value of the load current. Consideration must also be given to system start up and normal operational current surges, such as motor starting, to avoid nuisance fuse blowing under worst case conditions.

The available system fault current can be derived by dividing the source voltage by the series impedance (including source impedance) assuming a shorted load. This information, together with the selected fuse voltage and current rating, may be used to find the fuse peak let-through current, I_{plt}, and the total clearing I^2t, from the manufacturer's fuse tables. From these two parameters, the actual fuse clearing time (t_c) may be determined according to the following formula (based on a triangular pulse shape):

$$t_c = \frac{3\,(I^2t)}{I^2_{plt}}$$

Most semiconductor fuse manufacturers provide subcycle fuse-semiconductor coordination charts to simplify fuse selection. Fuse data is often presented with variables such as the fuse value and available system fault current integrated into a single log log graph (Fig. 9-18) for total clearing (let-through) I^2t versus total clearing time (t_c). With these coordinates, the SSR I^2t values can be plotted right onto the graph and a maximum fuse value readily selected. For example, the approximated I^2t plot shown for a 25-ampere SSR indicates that a 25-ampere fuse (of the type depicted) will protect the SSR with available fault currents up to 1000 amperes. Larger available fault currents require a lower current fuse or a higher current rated SSR.

A similar type of presentation (Fig. 9-19) shows the same variables included in a graph for peak let-through current (plt) versus total clearing time (t_c). In this case, the SSR surge current curve can be plotted directly onto the graph.

A third, more common subcycle fuse presentation, is a graph for total clearing (let-through) I^2t versus available fault current, which also includes fuse values and actual clearing times. Again, SSR I^2t values can be plotted onto the graph, as shown in Fig. 9-20.

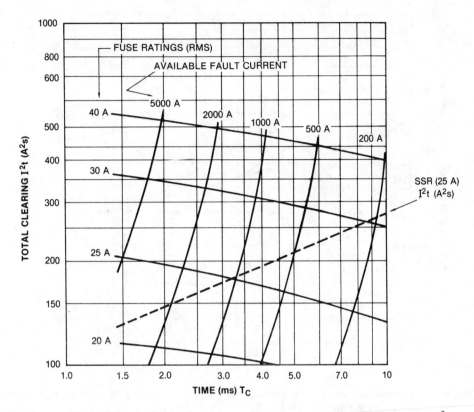

Fig. 9-18. *Typical fuse-semiconductor coordination curves for total clearing I²t versus total clearing time t_c.*

In practice, these fuse selection techniques are simple only if the designer has a thyristor data sheet at hand, since the two I^2t values necessary to make a plot, or subcycle surge ratings, are rarely if ever given on an SSR data sheet. Subcycle I^2t values, and a reasonable facsimile of surge current, can be computed using the $I^2\sqrt{t}$ formula and the given single cycle surge value, as previously mentioned. I^2t at 1.5 milliseconds is typically 50 to 60 percent less than the given 8.3-millisecond value. In the following example, the single cycle surge of a 25-ampere SSR is given as 250 amperes:

$$I^2\sqrt{t} = \frac{250^2}{2} \times \sqrt{.0083}$$

$$= 31250 \times .091$$

$$= 2844 \text{ A}^2\sqrt{\text{second}}$$

$$I^2t \text{ at time } t_x = I^2\sqrt{t} \times \sqrt{t_x}$$

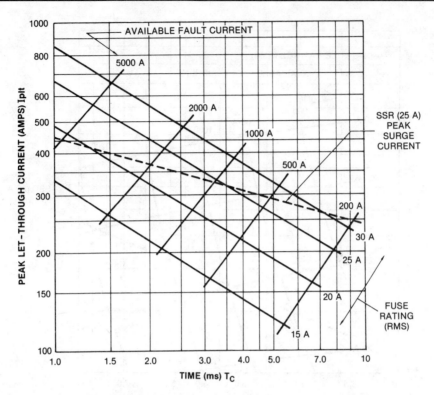

Fig. 9-19. *Typical fuse-semiconductor coordination curves for peak let-through current versus total clearing time t_c.*

Therefore:

$$I^2t \text{ (at } t_x = 1.5 \text{ ms)} = 2844 \times \sqrt{.0015}$$
$$= 2844 \times .0387$$
$$= 110 \text{ A}^2\text{s}$$
$$I^2t \text{ (at } t_x = 5 \text{ ms)} = 2844 \times \sqrt{.005}$$
$$= 2844 \times .0707$$
$$= 201 \text{ A}^2\text{s}$$

The I^2t values for 1.5 milliseconds and 5 milliseconds can now be inserted in the graph of Fig. 9-20, indicating that a 25-ampere fuse will protect the SSR if the available fault current is no greater than 550 amperes. With fault currents up to 3000 amperes, a 20-ampere fuse would be required, thus reducing the SSR switching capability. If operation at 25 amperes is necessary, a larger SSR with a higher I^2t, or the insertion of current-limiting resistance and/or reactive inductance, would be the solution.

SSR peak surge current values for the subcycle region can be estimated from the I^2t values using the formula:

$$I_{\text{pk surge}} = \sqrt{\frac{I^2t}{t_x} \times 2}$$

118

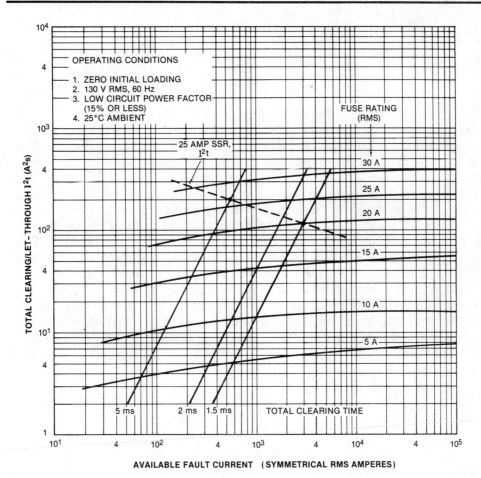

Fig. 9-20. *Typical fuse semiconductor coordination curves for total let-through I^2t versus maximum available fault current.*

e.g., if $I^2t = 110$, and $t_x = 1.5$ milliseconds

$$I_{pk\,surge} = \sqrt{\frac{110}{.0015} \times 2}$$

$$= 383 \text{ amperes}$$

This value, together with the given 8.3-millisecond value (250 amperes), can be inserted into a graph such as in Fig. 9-19 where peak let-through current (plt) is plotted against fuse clearing time.

When the current is known, the time element (t_x) in I^2t can be found from the following:

$$t_x = \frac{I^2t}{I^2_{rms}}$$

119

or

$$t_x = \frac{I^2t}{I^2_{pk}} \times 2$$

As a point of interest, the I^2t let-through value of a fuse is influenced by the inductance to resistance (L/R) ratio of the load circuit. The I^2t value rises with an increase in the L/R ratio, and a correction factor or a curve is usually given in the manufacturer's data to compensate for this. Similarly, fuse performance is affected by temperature, where minimum fusing current is reduced by an increasing ambient temperature. Again, compensating factors or curves are given by the fuse manufacturer.

Some SSR manufacturers simplify the fuse selection process by providing a table with appropriate fuse part numbers for each product. This is obviously the easiest and possibly the most reliable method, especially if both the fuse and the SSR suppliers are one and the same. However, the assumption cannot be broadly made that a fuse selected by one SSR manufacturer for its products will automatically be suitable for another, even with comparable ratings. The fuse may have been selected on the basis of empirical data and/or a higher surge capability than that given in the published ratings.

10

Input/Output Interface Modules for Microcomputers

The material in this chapter is based upon the author's original article entitled *Modular I/O Interface Designs for Micros*, published in Design News, November, 1980 by Cahners Publishing Co.

10.1 What Is an I/O Module?

The I/O module is a specialized type of solid-state relay designed specifically to meet the demands of the industrial control market. Solid-state relay technology has, in the I/O module, produced a rather timely ready-made tool for interfacing modern computerized machine and process controllers with their respective loads and sensing switches. Solid-state I/O modules are a relatively new innovation and continue to benefit from the advances made in SSR technology, namely, in the areas of circuit integration and automated assembly.

While a number of different I/O package configurations appeared initially, they are now reduced to a couple of basic types, together with a measure of standardization that is taking place in both physical and

121

electrical parameters among manufacturers. Figs. 10-1 and 10-2 illustrate the more common package styles. The modules in Fig. 10-2 are the more popular. They have plug-in replacements available from multiple sources, where even the part numbers are common among manufacturers, which is a first for SSRs.

Fig. 10-1. *Typical panel mount I/O modules with integral power terminals and LED status indicator. (Courtesy Teledyne Solid-State Products)*

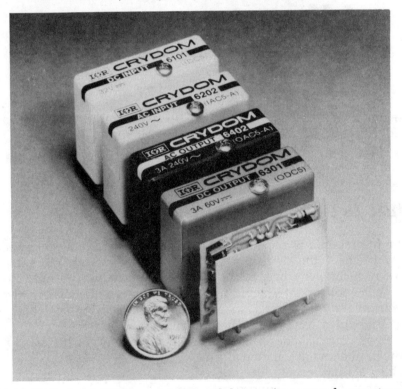

Fig. 10-2. *Typical panel/PCB mount I/O modules require external power terminals, status indicator, etc. (Courtesy IR Crydom)*

10.2 I/O Module Types

Most I/O modules are self-contained, and some include power terminals, LED indicator, internal transient suppression, and fuses (Fig. 10-1). They can be mounted individually by the user or plugged into a mounting base supplied by the manufacturer (holds up to 24 modules). The mounting base, which may also be used to accommodate power terminals and peripheral components, completes the wiring to these components, together with individual data lines and common bussing to the controller via a connector and single cable (Fig. 10-3). The pinout is generally compatible with a number of manufacturers' single board computers to facilitate the use of ribbon cable. Some manufacturers offer adapter cables to expand their application capability. Others offer remote control systems for communicating over long distances.

Functionally, there are four basic types of I/O modules: AC and DC inputs to logic, and logic to AC and DC outputs. Each module contains all the necessary circuitry to provide complete interface between controller and load without the need for further design or additional circuitry. To simplify assembly and field maintenance, modules are color coded and marked according to function.

All modules are photoisolated for electrical isolation (typically 1500 to 4000 volts RMS) between sensitive logic circuits and power lines, and have single form A (SPST) outputs. Multiple pole and special functions are not included since they are more easily configured in the processor logic, thus allowing for standardization and economies in the interface system.

A typical I/O modular system is shown schematically in Fig. 10-4, which, in addition to showing logic flow, illustrates the individual module circuit elements.

10.3 Input Modules

Input modules are unique devices in that they perform a reverse relay function in returning information from field contacts and loads to the computing source (Fig. 10-4 parts A and B). Through debounce and noise-suppression circuitry, the modules convert power-level field control signals into electrically clean logic-level signals suitable for microprocessor and controller inputs.

Input circuits include rectification and filtering of AC signals, and current limiting for DC signals to permit a wide range of control voltages and improved coupler lifetime. After being conditioned, the signals are photocoupled through driving stages to an open-collector logic output.

The open-collector output can usually sink sufficient current (typically 25 milliamperes) to allow direct interface with any logic family operating within the voltage range of the output transistor (typically 0.2 to 30 volts DC). A third terminal in the output provides access to the

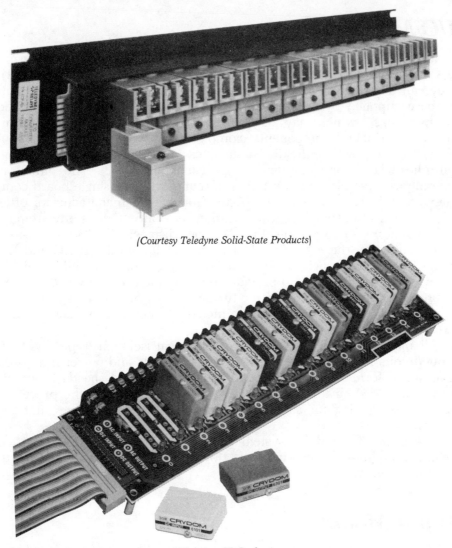

(Courtesy Teledyne Solid-State Products)

(Courtesy IR Crydom)

Fig. 10-3. *I/O boards simplify interface to microprocessor systems.*

logic supply for biasing purposes and for the open-collector output function. The LED status indicator, which is also in series with the logic supply, can either be an integral part of the module as shown, or installed on the module mounting board.

The input signals, derived from limit switches, loads, thermostats, pressure transducers, etc., are used by the computing logic in its decision-making process. (In a programmable controller, this process might determine a sequence, timing, or make accept/reject decisions, from which predetermined commands that ultimately drive output modules can be generated.)

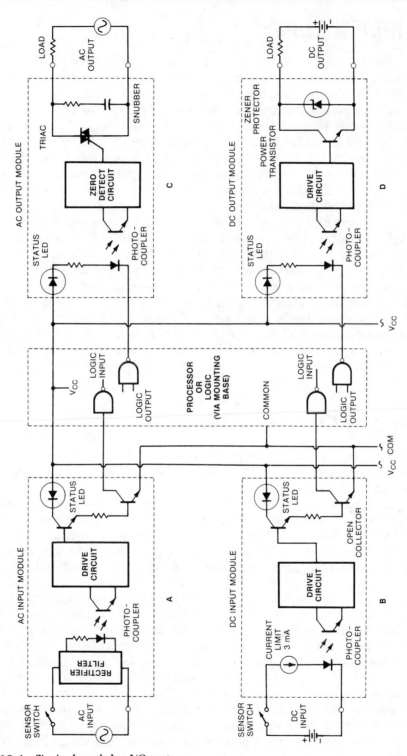

Fig. 10-4. *Typical modular I/O system.*

10.4 Output Modules

Output modules (Fig. 10-4 parts C and D) are similar in function to conventional AC and DC solid-state relays. The types that are generally available can handle most power-switching applications found in machine and process control, up to 4 amperes at 40°C, AC or DC.

Most existing output modules can be driven from TTL logic (in the sink mode), while newer buffered versions have the ability to be driven directly from CMOS or NMOS logic. Unlike input modules, standard output modules have four terminals (no logic supply requirements), with the internal or external status indicator driven by the control signal itself (usually in series with the positive leg).

Driven from low-level logic command signals, output modules are designed to control power to loads such as motors, valves, and heaters, and generally include a high degree of noise rejection and transient-voltage suppression. AC versions also include zero voltage turn-on to reduce surges and the generation of electromagnetic interference, as well as zero current turn-off (an inherent characteristic of triacs), an advantage when switching inductive loads. Snubbers and filter components provide improved commutation and dv/dt performance and further improve immunity to transients and noise. DC outputs usually include a zener diode for transient-voltage protection.

10.5 Buffered Output Modules

Buffered output modules contain additional internal amplification to reduce drive requirements to a level suitable for the MOS devices used in many microprocessor systems.

To accommodate the different turn-on states of microprocessors, the buffered modules are available in both inverting and noninverting versions. By definition, a noninverting buffered output module turns on when the input is held in the low state (logic 0), as do standard modules. An inverting module conversely turns on when the input is held high (logic 1).

For total MOS compatibility, inverting buffered modules should operate with an external current source of 100 microamperes maximum, while noninverting modules should operate with an external sink current of 200 microamperes maximum. These parameters meet the requirements of all CMOS and NMOS logic families; thus, existing microprocessor boards utilizing these logic systems do not require additional interfacing components. This higher input sensitivity is usually attained by the addition of extra gain stages, as illustrated in the simplified schematics of Fig. 10-5. Note that a buffered output module has a fifth terminal to provide power to the photocoupler, and a LED status indicator if required.

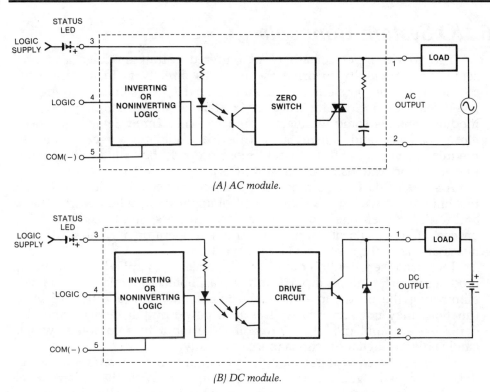

(A) AC module.

(B) DC module.

Fig. 10-5. *Buffered output module schematics.*

The inherently high gain and resulting increased sensitivity to noise of buffered modules has been the cause of some concern. The higher sensitivity can be controlled by tailoring the module input impedance to suit the application. A convenient way to do this is by lowering the value of the pull-up resistor, located on the module mounting board, to suit the power capability of the driving gate. Since the noise required to turn on a module is also a function of the driver output impedance, it is the power capability of the driver that ultimately determines the system noise immunity.

As a point of interest, it would appear that the maximum input current of a standard output module, 15 milliamperes at 5 volts, is far too high to be driven by even the highest rated microprocessor or peripheral interface adapter (PIA) outputs. The 15 milliamperes is a "forced" value established by an input impedance of 240 to 330 ohms and is designed to illuminate the visible status LED. Without the LED the actual module drive requirement is closer to 2 milliamperes at 3 volts (across the input pins), which is within the realm of possibility for some CMOS outputs. The manufacturer should be checked on this point, since operating a standard module close to its operating threshold could result in a high susceptibility to noise. The stability of the lower operating threshold over the temperature range is also a consideration.

10.6 I/O System Concepts

The greater input sensitivity of buffered output modules also presents the opportunity to drive the modules directly from a transducer source, with the possibility of including AND/OR, latching, timing, or level detection logic. This suggests a stand-alone situation where the module contains some intelligence and becomes three devices in one— *input, computer,* and *output.* Such a device would no longer be an I/O module or even an SSR, but could more properly be called a logic module. Some applications for intelligent I/O's are given in Table 10-1.

A series of these logic modules placed in strategic locations around a machine might well suit a simple control application, where wiring back and forth between one central controller, detectors, and loads would not be cost effective. Local control by a human operator may be all that is necessary, as in the case of a vending machine.

Increased sensitivity could also be usefully applied to an input module, where a direct input from, say a photocell, could be accepted, eliminating the need for additional components. The addition of logic functions may not serve as well as in a power-output device, since by definition an input module is providing the input to a computer which can handle the digital functions more efficiently.

A certain breed of input module is emerging that converts analog signals into digital signals usable by the computer. With the voltage-level sensor, a comparator receives an analog transducer signal and compares it with a built-in reference voltage. The output state of the module changes abruptly when the analog signal varies from the predetermined reference point. In a more advanced version, the trip point might be variable, with the setting determined by a control signal from the computer.

In another type of analog converter module, the varying characteristics of a transducer output (such as a strain gage) are amplified, then digitized incrementally, possibly by means of an A/D converter. In this way, a computer is fed analog information in an acceptable form on a continuing basis. The accuracy or resolution of this information depends largely on the number of increments used to define a particular characteristic, as in the case of plots describing a curve on a graph.

In addition to translating the analog information into digital form, the module may also include an operational amplifier that can be controlled by the computer to linearize or modify the signal before conversion. In short, the intelligent I/O may be used to house all the functions that are more difficult to implement in the computer.

When large numbers of I/O modules are used in a system, the wiring alone can become a very large problem. Some computing systems use multiplexing to address the individual I/O's and thereby reduce the amount of wiring necessary. One such system requires that each module contain a refreshable memory which, in effect, is a latch combined with a time-delay on dropout. The address is accomplished by scanning in an XY manner, with data lines common to modules in vertical columns,

Table 10-1. *Applications for Intelligent I/O Logic Modules*

BUILT-IN LOGIC		I/O MODULE CONTROL FUNCTIONS
	AND/OR	Local enable/inhibit warning, signal intercept directly at site of machine being controlled.
	Latch	Pulse driven or used for pulse/event detection and memory. Local alternate action function. Facilitates one form of multiplexing.
	Time Delay: On-Operate One-Shot/Interval On-Release	Programmable at site or remotely by computer. Prevent make-before-break. Pulse stretching function. Facilitates multiplexing.
	Comparator	Threshold detector—set point on analog signal. Level sensor over/under voltage, current frequency, phase, limit switching. Can be programmable at site or remotely by computer.
	Operational Amplifier	Amplify analog signal. Control, limit, modulate, shape, filter signal characteristic. Can be programmable at site or remotely by computer.
	Encoder/ Decoder	Minimal wiring to and from remote locations. Remote-enable/inhibit/program of I/O modules.
	Analog to digital converter	Translate analog information into digital form usable by computer.

while each row is horizontally activated by sequentially applying V_{cc} (5 volts). The scan sequence is rotated at a rate that maintains steady-state logic levels.

The foregoing system depends entirely on specially designed I/O modules for its operation. A system that contains the necessary logic for multiplexing in the module mounting base can make use of standard multiple source modules. In this case, the mounting base becomes specialized and unique to that system, but is less likely to change and is more economical.

The problem of addressing large numbers of I/O modules with fewer data lines can be dealt with effectively if individual modules, or groups of modules, each contain a specific address code. Besides communicating in standard binary computer language, the modules no longer need to be clustered together in one place. With individual identity addressable over a few lines, output modules with some intelligence (such as the logic modules mentioned above) become stand-alone controllers that take orders from a central computer or microprocessor.

The addressable output module would be very effective in the control of remotely located power loads in such application areas as energy management. Input modules, also remotely located with the same capabilities, may be used to return load-related information, such as activation confirmation, back to the computer.

It may be in this form that the I/O module will finally reach the home appliance and consumer markets. It could provide an easy to install interface between the home computer and the numerous domestic loads—pool, garden and solar energy equipment—where automatic control and power conservation would be beneficial.

The four channels of I/O in one package shown in Fig. 10-6 illustrates a more recent innovation in the field of I/O modules. The condensed packaging and mounting board stacking permits the use of smaller enclosures.

Fig. 10-6. *Quad-Pak™, high density I/O module.* *(Courtesy Opto-22)*

11

SSR
Applications

The diagrams in this chapter are conceptual illustrations of just a few typical SSR applications. They are intended as design guides to steer the user in the right direction and to stimulate further design ideas. Some of the diagrams provide problem solving or circuit protection, and others enhance relay operation. They are also a practical extension of the many discussions contained in the previous chapters.

SSRs cannot always be applied in exactly the same way as EMRs and when such is the case, cautionary notes advise the user of possible pitfalls and suggested ways of overcoming them. It is assumed that the reader has a working knowledge of electronics and can fill in the gaps, where necessary, to complete the designs. In some cases, particular functions such as time delay or SPST switching, have been repeated in different applications with slight variations. This was done deliberately to illustrate the different approaches which are appropriate for the associated application, but not necessarily tied to it. Also, where fusing is suggested as insurance against shorts or to protect the wiring, other circuits may also benefit from fusing, even though its use may be less crucial. Likewise, when the faster more expensive semiconductor fuses are suggested to preserve the SSRs, other circuits may be similarly protected wherever the additional cost is justified, or SSRs are difficult to replace.

11.1 Complementary Power Switching (Fig. 11-1)

Two SSRs (AC or DC) can be used to switch power between two loads in complementary fashion, simulating a single-pole double-throw action. Due to the possibility of overlap (make-before-break) the power source should be capable of supporting both loads momentarily. No attempt should be made to switch a single load between two power sources.

Typical loads would include lamps, valves, and solenoids for GO/NO GO or high-low applications. For level detection, fairly precise current or voltage triggering thresholds may be established by using the more sensitive, buffered AC output (I/O) modules in the position of SSR1.

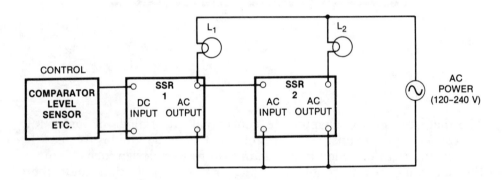

Fig. 11-1. *Circuit illustrating complementary power switching.*

11.2 Latching SSR (Fig. 11-2)

Momentary push-button control allows the SSR to self-latch for on-off, stop-start operations. It may be similarly configured for DC in/DC out type SSRs.

Resistor R1 (10,000 ohms) is required to prevent line short only if alternate (N O) switch is used.

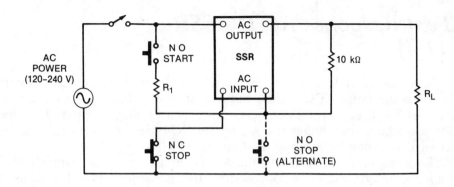

Fig. 11-2. *Latching SSR circuit.*

11.3 Latching SSR with Short-Circuit Protection (Fig. 11-3)

Push-button control as in Fig. 11-2 but R2 is tailored to limit the load shorting current to SSR surge rating (for turn-off time), thus preserving SSR while control signal is removed. Latching characteristic permits lock-out until the circuit is reset.

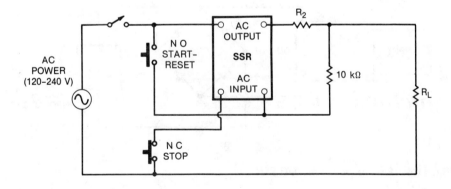

Fig. 11-3. *Latching SSR with short-circuit protection.*

11.4 Fast Response AC/DC SSR (Fig. 11-4)

Two configurations of DC SSRs are used in bridge circuits to switch AC or nonpolarized DC. While bidirectional switching more closely resembles EMR mechanical contacts, the forward voltage drop and dissipation are higher than either the AC or the DC types because of the added series diodes, one for A of Fig. 11-4 and two for B of Fig. 11-4.

Features of the circuit include a fast on-off response, typically less than 250 microseconds as opposed to the possible 8.3 milliseconds of thyristor types. These configurations are also immune to dv/dt since they will not latch on like a thyristor in response to a rapidly rising voltage.

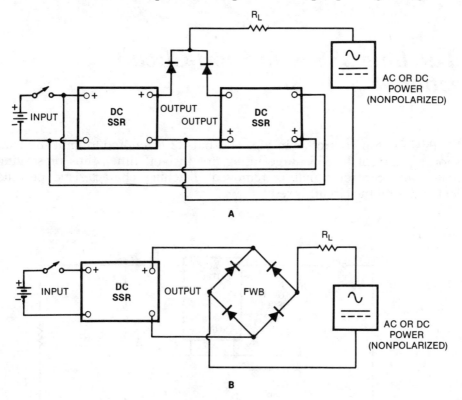

Fig. 11-4. *Fast response AC/DC SSR.*

11.5 SSR/EMR Hybrid Relay (Fig. 11-5)

This configuration possibly represents the best of both worlds, by combining the SSR soft, bounce-free zero turn on-off with the low contact drop (dissipation) of EMR contacts. When S1 closes, C1 rapidly charges through R1 and the SSR turns on, followed by the slower contact closure of K1. When S1 opens, the RC network of C1, R2, plus the SSR input impedance, holds the SSR on until the contacts of K1 open, thus permitting a zero current turn-off.

This circuit can be similarly configured for DC output SSRs or AC input SSRs and EMRs. In considering relative operate and drop-out times, it may be found that some AC input relay combinations will operate in the proper sequence without the delay network.

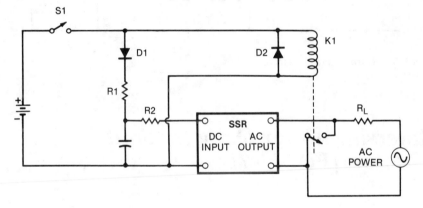

Fig. 11-5. *SSR/EMR hybrid relay.*

11.6 Motor Starter Switch (Fig. 11-6)

Initial locked rotor current flowing through R1 creates a voltage that, when rectified and filtered, turns on the SSR which in turn activates the start winding. As the motor comes to speed, the voltage across R1 is reduced until the start winding is deenergized.

The SSR should have a voltage rating approximately twice that of the applied line to withstand overvoltage generated by the circuit LC.

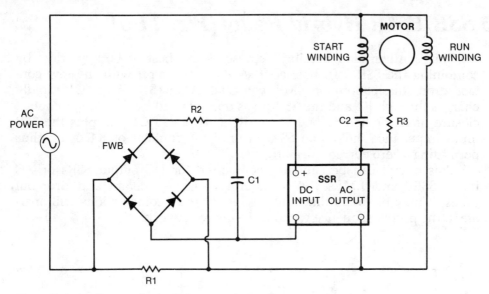

Fig. 11-6. *Motor starting switch.*

11.7 Reversing Motor Drive for Split Phase Motors (Fig. 11-7)

Two relays are used in this motor-reversing circuit. The time-delay on-operate is preferred (but not essential) to avoid make-before-break and high-voltage peaks at the moment of reversal. To withstand these peaks, the relays should be rated at twice the applied line voltage.

Resistor R1 should be sized to limit the current to something less than the peak one cycle surge-current rating of the SSRs. Should the capacitor C1 be inaccessible, the resistor value can be split and placed in series with each relay output.

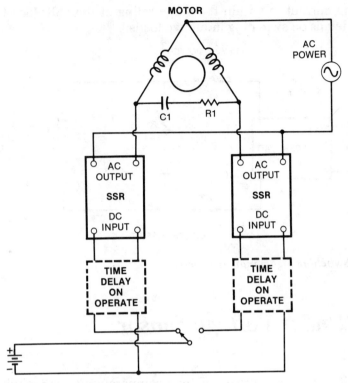

Fig. 11-7. *Motor-reversing circuit.*

11.8 Switching Highly Inductive Loads
(Fig. 11-8)

Most SSR manufacturers guarantee inductive load operation with a minimum power factor of 0.5. If erratic operation is experienced with lower power factors, it is possible that the SSR is not reaching latching current during the interval that the gating signal is present.

One corrective measure is to add an RC network across the load to increase its power factor. Typical values that have frequently proven successful are R1=150 ohms and C1=0.5 microfarad. The resistor is necessary to prevent the instantaneous charging current of C1 from damaging the SSR.

Another solution would be the use of a nonzero (random) turn-on SSR, which would eliminate the timing restrictions imposed by the zero switching window, at the sacrifice of the initial zero voltage turn-on.

Heavy current well in excess of ten times steady-state current can sometimes occur with inductive loads at initial turn-on, or due to erratic operation of the SSR. In this event, resistor R2 should be placed in series to limit the current to within the surge rating of the SSR. (See Chapter 8 for more details on switching inductive loads.)

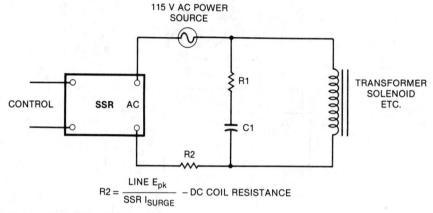

$$R2 = \frac{LINE\ E_{pk}}{SSR\ I_{SURGE}} - DC\ COIL\ RESISTANCE$$

Fig. 11-8. *Switching highly inductive loads.*

11.9 Over/Under Voltage Sensor (Fig. 11-9)

The MC3425 is comprised of four comparators together with all the necessary functions for overvoltage and undervoltage sensing, including a built-in common 2.5 volt reference. The two sensing channels are shown reversed in function to simplify the SSR drive, which is on when the line voltage is within the prescribed limits set up by the potentiometers.

The time delay for each channel is determined by capacitors connected to pins 2 and 5 which provide a constant current source of typically 200 microamperes. (The overvoltage input comparator, pin 4, has a feedback-activated 12.5 microampere current sink for programming hysteresis, determined by input source impedance; see manufacturer's data for more details.)

With the input of the sensor shown connected between phases, SSRs in the output, with inputs paralleled, can be used to shut down power to a sensitive three-phase load when line voltage goes out of tolerance. This circuit may also be programmed as a frequency-sensitive switch.

Using only the undervoltage channel (pin 3) for overvoltage and "source" driving the SSR from pin 1 to ground (−), the output can be used to "crowbar" its own supply. Without the AC bridge and filter, the MC3425 operates directly from DC, and as the front-end of an SSR, it can be used to perform many useful level-detection functions.

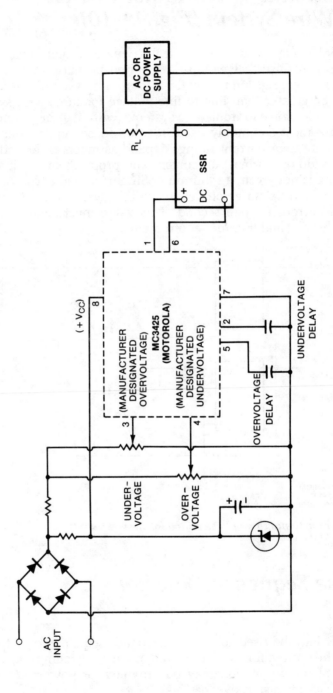

Fig. 11-9. *An undervoltage-overvoltage sensor.*

11.10 Functional Three-Phase Switch for Three-Wire System (Fig. 11-10)

Two SSRs may be used to control a wye or a delta load in a three-wire system. A third SSR would be required in phase C if the center of the wye load were grounded, as in a four-wire system. SSR voltage ratings must be greater than line to line voltage for three-wire systems and line to ground voltage for four-wire systems (with neutral ground).

SSRs are most commonly used in three-phase applications to control motors, where their current ratings depend as much on locked rotor current as they do on normal run current and proper heat sinking. Where a motor rating is not given, a minimum SSR current value can be estimated from manufacturers' surge curves, using the general rule of six times the motor run current for one second. This value must also be commensurate with thermal and lifetime requirements.

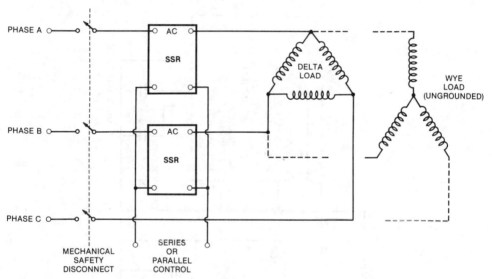

Fig. 11-10. *Functional three-phase switch for three-wire system.*

11.11 Phase-Sequence Detector (Fig. 11-11)

An SSR may be used to detect the phase sequence of a three-phase system, while providing isolation and power switching capability. An indication of motor rotation may be provided, or power interrupted, to prevent damage to a sensitive load.

The SSR should have a control sensitivity of less than 2 milliamperes, and a low input impedance compared to the R1, R2, C1 combination values. The vectorially combined currents of I1 and I2 provide a control signal of approximately 2 milliamperes, driving the SSR on when phase A is leading phase B. When they are reversed, with phase A lagging phase B, currents I1 and I2 will be in opposite phase and cancel, causing the SSR to remain off until the sequence is reversed.

Values may be adjusted to provide sufficient drive for three SSRs, and thus directly control power to a three-phase load.

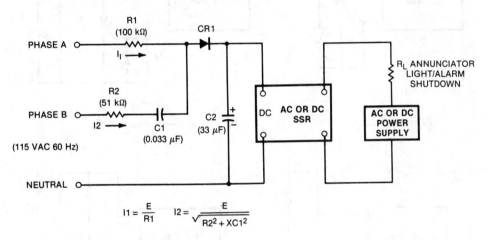

$$I1 = \frac{E}{R1} \qquad I2 = \frac{E}{\sqrt{R2^2 + XC1^2}}$$

Fig. 11-11. *A phase-sequence detector.*

11.12 Driving High Powered Thyristor and Bipolar Devices (Fig. 11-12)

If available SSRs are inadequately rated, SSRs may also be used to drive higher powered discrete devices as shown in the illustrations.

As a cautionary measure, series resistor R_X is included in the circuits of Figs. 11-12A and B to limit the drive current to the surge rating of the SSRs. R_Y must be used in all cases to prevent the SSR leakage from turning on the power device in its off-state. The diodes in Fig. 11-12B will prevent damage to certain types of SCR (nonshorted gate types).

With the simple drive method of Fig. 11-12C, the dissipation in the power device can be as high as 2 watts per ampere. By providing separate bias by means of R_Z in Fig. 11-12D, part of the dissipation is transferred to R_Z, but dissipation in the power device may be reduced by a factor of ten.

In Figs. 11-12A, B, and C, load R_L may be situated in either leg of the power source. In Fig. 11-12D, the load is dedicated to the positive leg of the supply, but may be reorganized for a ground $(-)$ referenced load with a PNP transistor.

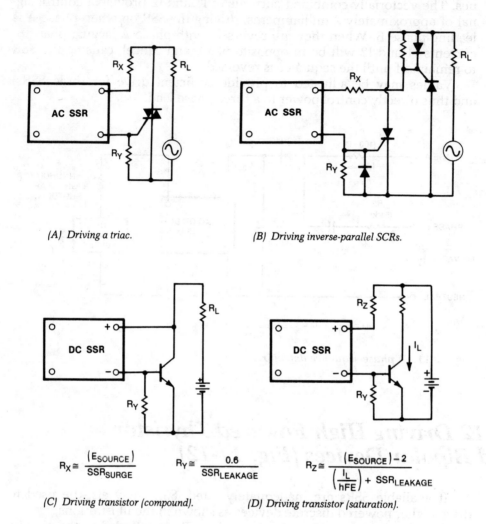

(A) Driving a triac.   *(B) Driving inverse-parallel SCRs.*

$$R_X \cong \frac{(E_{SOURCE})}{SSR_{SURGE}} \qquad R_Y \cong \frac{0.6}{SSR_{LEAKAGE}} \qquad R_Z \cong \frac{(E_{SOURCE}) - 2}{\left(\frac{I_L}{hFE}\right) + SSR_{LEAKAGE}}$$

(C) Driving transistor (compound).   *(D) Driving transistor (saturation).*

Fig. 11-12. *Driving high powered thyristor and bipolar devices.*

11.13 DPDT Switch from Single Transistor Source (Fig. 11-13)

A number of multipole configurations can be organized from a single transistor drive such as that shown.

For a 5-volt DC control power source, the dual inputs shown in series should be in parallel to reduce the turn-on threshold from a typical 6 volts to 3 volts. The input current (I_{F1} and I_{F2}) would then become double that of the series configuration shown.

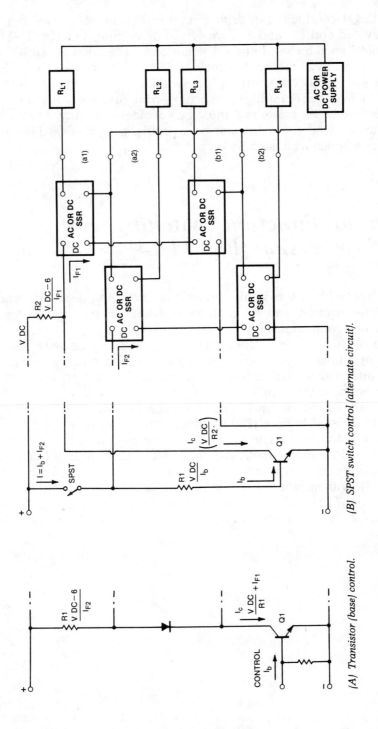

Fig. 11-13. *DPDT switch from single transistor source.*

With Q1 conducting in drive circuit in Fig. 11-13A, loads R_{L1} and R_{L3} are activated and R_{L2} and R_{L4} are off. In drive circuit of Fig. 11-13B, with the control switch closed and Q1 conducting, the opposite would be true. As an alternative, the switch may be placed in the base of Q1 in Fig. 11-13A where only base current I_b is switched and its function inverted.

As in many SSR multiple switching configurations, the possibility of make-before-break exists and must be considered in any design. A possible solution, where overlap can be a problem, is often a time delay (on operate) in series with each SSR input.

11.14 Special Function Switching with Bounce Suppression (Fig. 11-14)

A variety of SSR switching functions, including debounce and latching where required, can be implemented with a few "spare" logic gates and discrete components.

The illustrations are conceptual and inputs can be with switches as shown or similarly oriented logic signals. The debounce feature is primarily intended for fast acting DC SSRs, since the thyristor in AC types already provides this function.

All circuits operate on the leading edge (switch closure), with bounce immunity related to time (100 milliseconds) by means of an RC network, in Figs. 11-14A and C. The network is not required in Fig. 11-14B by virtue of the dual momentary switches. All circuits also provide optional inverted outputs which may be used for the SPDT switching function (Fig. 11-1), with appropriate precautions.

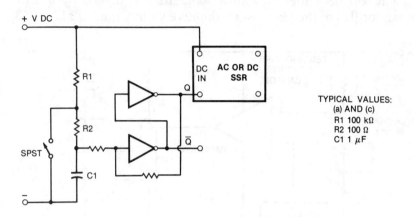

(A) Two inverters form Schmitt trigger for switch debounce.

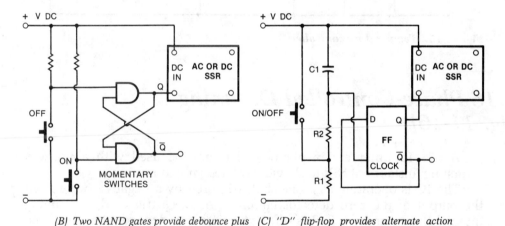

(B) Two NAND gates provide debounce plus latching function.

(C) "D" flip-flop provides alternate action switch with debounce and latch.

Fig. 11-14. *Special function switching with bounce suppression.*

11.15 Temperature Controller (Fig. 11-15)

The SSR is ideal for use in temperature control systems such as that shown, or to replace mechanical contacts in existing systems. The zero switching characteristics virtually eliminate contact noise and wear, thus improving reliability.

A very simple temperature controller can be implemented with a thermostat controlling an SSR or a thermistor bridge at the input of a sensitive SSR such as a buffered AC output (I/O) module. However, greater precision can be achieved with a standard SSR driven by a differential comparator (IC) of the low cost automotive variety (i.e., 1/2 LM2903).

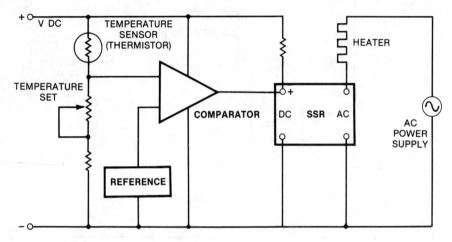

Fig. 11-15. *Temperature controller.*

11.16 Phase-Controlled Dimming (Fig. 11-16)

A 555 IC timer and a photocoupler may be used with a nonzero switching (instant-on) SSR to provide isolated lamp dimming.

The IC is operating as a one-shot, triggered by a negative pulse from the output of the zero detector circuit (Q1). Once triggered, the timing interval begins and the SSR is off. Upon time-out, dependent on the time constant of R1C1, IC output (pin 3) goes low and the SSR turns on for the balance of the half cycle. Simultaneously, C1 is discharged through a transistor in the IC (pin 7), and the process repeats every half cycle.

The phase angle firing point is independent of DC control voltage. However, at the higher DC voltages and shorter firing angles (full-on), the IC may overdissipate due to the repetitive discharge of C1, and possibly require a small heat sink.

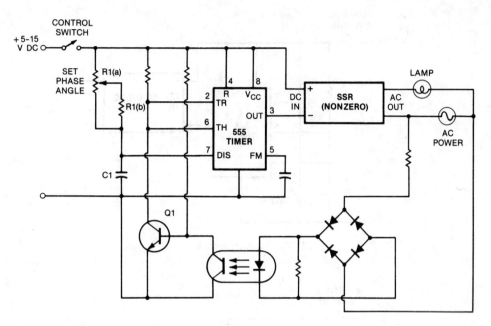

Fig. 11-16. *Phase-controlled dimming.*

11.17 Time Delay/Multivibrator (Fig. 11-17)

The 555 LC timer, in conjunction with very few additional components, can provide time delay and multivibrator (flasher) functions for the SSR.

With SW2 closed (Fig. 11-17A) the circuit will operate as a time-delay on-operate when SW1 closes, as determined by the time constant of R1 and C1. Alternatively, with SW1 closed the SSR will turn off for the same period with each momentary closure of SW2, similar to the operation of Q1 in Fig. 11-16. The above switching functions may be inverted by placing the SSR in "source" driven mode, as in Fig. 11-17B.

In Fig. 11-17B, SW1 closure causes capacitor C1 to charge through R1 and R2 for the on period and discharge through R2 only for the off period. The circuit automatically retriggers itself each cycle forming a multivibrator, with the timing interval determined by the ratio of the two resistors. For a duty cycle of less than 50 percent, it is necessary to add diode CR1, modifying the charge path for C1 through the diode to R1 only.

The timing intervals in both circuits are independent of the DC control voltage.

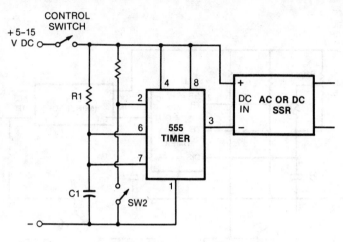

(A) Time delay (monostable).

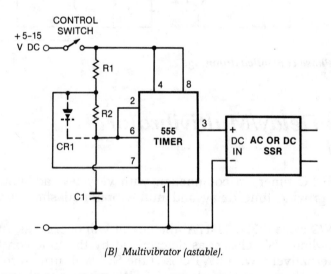

(B) Multivibrator (astable).

Fig. 11-17. *Time delay/multivibrator.*

11.18 Hazardous Applications

When applying SSRs in certain high power switch applications, the designer should be fully aware of all SSR and system characteristics that might, in combination, produce a hazardous situation for the user or the equipment. For example, the natural SSR sensitivity to overvoltage transients and fast rising voltages (dv/dt) can produce a brief "on" condition that, together with the tendency to make-before-break, might prove catastrophic when switching between power lines or supplies. However, with appropriate circuitry and protective measures, an acceptable design can usually be worked out, assuming that properly rated SSRs are available. The circuits illustrated in Figs. 11-18 through 11-22 are considered hazardous and require special care in their implementation.

The careful use of fusing, to provide back-up protection without nuisance fuse blowing, is a wise precaution in these applications to ensure complete safety from fire hazards. Of those applications with a high order of risk, motor reversal on a three-phase system probably ranks among the highest. The limited lifetime and electrical noise produced by contactors has always been a problem in this type of application, making the excellent lifetime properties of SSRs particularly attractive.

Possibly the most reliable design, wherever practical, is still with the actual line reversal accomplished by mechanical switches or EMRs, with the power off and the motor stationary, while the start/stop operation is performed by SSRs. In such a configuration, the possibility of a line to line short is highly unlikely and the SSRs would provide arc-free on/off switching. Where this is not possible, the all solid-state relay circuit of Fig. 11-18 will apply.

In this circuit, due to the nature of SSRs, there is the possibility of two relays turning on simultaneously causing a line to line short; the design therefore should be capable of supporting such an event without failure. The precautions shown are mandatory in order to avert the potentially dangerous circumstances of a high energy short circuit. The practicality of the circuit is limited by the availability of SSRs with sufficiently high ratings to meet the suggested criteria, while permitting normal operation of the motor.

The circuit of Fig. 11-19 shows examples of SSRs being used to switch two power supplies, possibly at different potentials, into the same load. The risks again are high because a dual "on" condition can cause extremely high currents to flow between supplies. Suggested protective measures are also similar to those of Fig. 11-18. The drive circuits in each case are also similar in that their primary purpose is to prevent overlap (make-before-break).

The DC reversing motor drive of Fig. 11-20 uses four SSRs in pairs to reverse the polarity of a single power supply across the motor. The circuit of Fig. 11-19 can do the same thing with only two SSRs, but requires two supplies in tandem to do it. The risk of a dead short again exists, hence the same precautionary measures are suggested.

A somewhat different situation is found in the paralleling of SSRs in Fig. 11-21, where the risks apply mainly to the SSRs and their ability to perform reliably. The degree of success of this technique varies widely, depending on the SSR type. SSRs that use power (MOSFET) field effect devices in their outputs can be easily paralleled since their positive temperature coefficient forces current sharing without the use of power-robbing balancing resistors. SSRs using standard bipolar and thyristor devices, on the other hand, do require balancing resistors as illustrated to prevent current hogging, particularly where a derating factor of less than 20 percent is desired.

A major difficulty exists with the paralleling of thyristors where parameters over which the user has no control can prevent turn-on of one SSR causing destruction of the other. All thyristors have a voltage threshold higher than their on state that must be reached before turn-on can occur. This means that one device with a low threshold may turn on first, clamping off the applied voltage to its neighbor, thus preventing it from turning on. The varying propagation (delay) times of the command signal through the SSR also contribute to this problem.

The advisability of paralleling thyristor type SSRs must therefore be questioned. Balancing resistors will improve the situation somewhat and the use of balancing inductors improves it still further, but the risk is still there. One technique that is cumbersome, but has been successful, is to phase-fire nonzero SSRs well above the peak repetitive on voltage (PROV) at each half cycle, beyond the point where prefiring can occur.

In practice, if a combination of two or more SSRs is found to operate properly under all combinations of input, output, and temperature variations, the chances are good that it will continue to do so. However, because of the possible catastrophic consequences should one SSR fail, a thermal shutdown system might also be considered as back-up insurance.

The hazard in the transformer tap switching application in Fig. 11-22 is to the transformer itself, where a dual "on" condition could cause a winding to burn out. Again, the judicious use of fusing can prevent such an occurrence.

11.19 Three-Phase Motor Reversal (Fig. 11-18)

Four AC SSRs can provide a reversing function for a three-phase motor, using the drive logic suggested. The half cycle time delay before enabling the drive, in either direction, prevents make-before-break which would result in a line to line short. Two opposing SSRs (Nos 1 and 4 or 2 and 3) could still mistrigger simultaneously due to dv/dt or high-voltage transients; therefore, resistors R1 through R4 are inserted to limit the resultant surge current. The sum of any two resistors plus the source impedance should limit the shorting current to less than the peak one cycle surge rating of each relay.

Semiconductor-type fuses should be chosen to permit such a condition for one cycle and open as soon as possible thereafter. SSRs should have a transient (blocking) rating equal to twice the line to line voltage to withstand the combined line and back EMF generated at the moment of reversal.

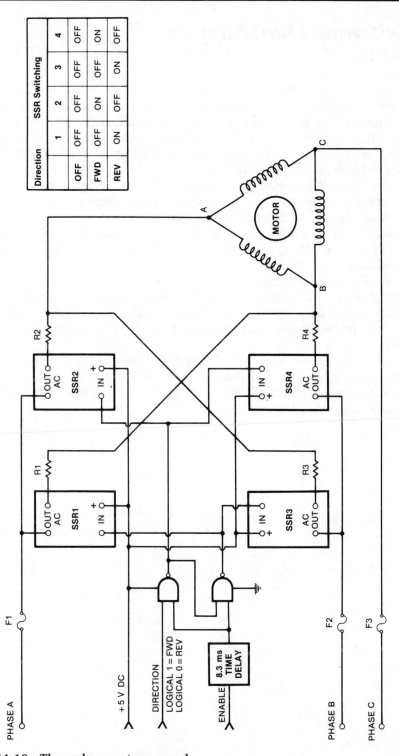

Direction		SSR Switching			
		1	2	3	4
OFF	OFF	OFF	OFF	OFF	OFF
FWD	OFF	OFF	ON	OFF	ON
REV	ON	ON	OFF	ON	OFF

Fig. 11-18. *Three-phase motor reversal.*

11.20 Switching Dual Supplies
(Fig. 11-19)

Switching two supplies as alternates into the same load is commonly required for auxiliary power switching, level switching, and motor reversal. Because of the danger of a dual SSR on condition causing heavy currents to flow between supplies, a simple break-before-make circuit is included. The turn-on delay provided by R1 and C1 should be longer than the SSR maximum turn-off time. Schmitt type inverting logic gates are recommended to avoid transitional problems. Fusing in the output will protect the wiring in all cases, should a malfunction occur.

In A of Fig. 11-19, R2 and R3 are considered necessary for AC circuits and are sized to limit the shorting current to the one cycle surge rating of the SSRs. Blocking voltages for each SSR must be at least twice the supply voltage.

In B of Fig. 11-19, the SSRs must withstand reverse voltage in the "blocking" mode. If the SSRs contain internal diodes or zeners, an external diode should be installed in series with each leg to block reverse conduction. A freewheeling arc-suppression diode across the load is also advised.

In C of Fig. 11-19, with a split supply (push-pull), the SSR blocking voltage should exceed the sum of the two supplies. In each case, an internal (or external) reverse diode across one relay output acts as an arc-suppression diode for the other.

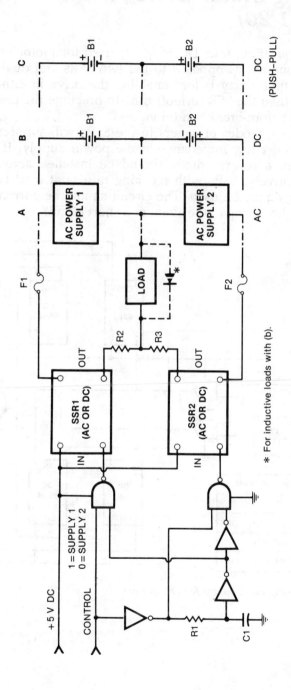

Fig. 11-19. *Switching dual supplies.*

11.21 Reversing Motor Drive for DC Motors (Fig. 11-20)

In this configuration, four DC SSRs are used for motor reversal from a single power supply, as opposed to the two SSRs and dual supplies of Fig. 11-19. The time delay before enabling the drive in either direction must be greater than the SSR turn-off time to preclude the possibility of a hazardous make-before-break condition.

Internal reverse diodes or zeners in the SSRs will suppress inductive transients across the low impedance of the power supply. If no internal suppressors exist, a reverse diode should be installed across each SSR output or alternatively, SSRs with blocking ratings at least twice that of the supply voltage may be used. The circuit should be current limited or fused to protect the wiring in the event of a short circuit.

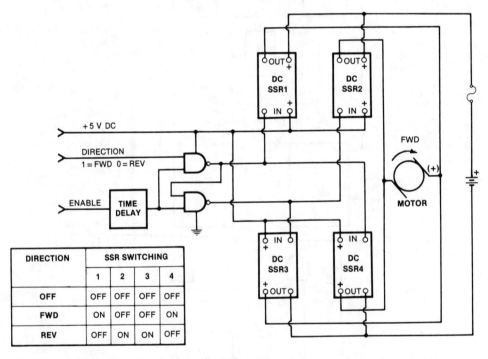

DIRECTION	SSR SWITCHING			
	1	2	3	4
OFF	OFF	OFF	OFF	OFF
FWD	ON	OFF	OFF	ON
REV	OFF	ON	ON	OFF

Fig. 11-20. *Reversing motor drive for DC motors.*

11.22 Paralleling SSRs (Fig. 11-21)

SSRs with MOSFET outputs are self balancing and easily paralleled, whereas most others with bipolar or thyristor outputs require special attention. Ideally, the forward voltage drops should be matched to achieve thermal balance and lowest dissipation; alternatively, balancing resistors (R_X) are used to force current sharing as shown. For example, with 40 amperes allowed through SSR1, SSR2 must carry 32 amperes:

Assuming V1 = 1.3 volts and V2 = 1.5 volts (worst case)

$$R_x = \frac{\Delta V}{\Delta I}$$

or

$$\frac{V2 - V1}{I1 - I2} = \frac{1.5 - 1.3}{40 - 32}$$

$$= 0.025 \text{ ohm}$$

thus producing a total voltage drop of 2.3 volts.

For zero voltage turn-on thyristor types, either one of the SSRs must be capable of handling the initial full load surge alone because of a possible half cycle mismatch. Thyristor SSRs have additional turn-on problems that can prevent paralleling. See text, Section 11.18.

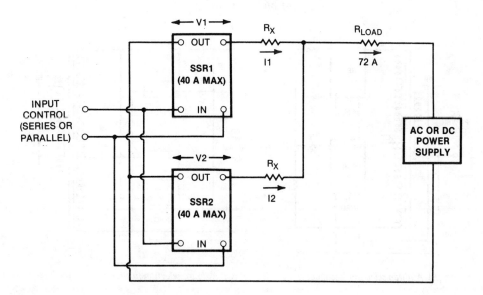

Fig. 11-21. Paralleling SSRs.

11.23 Transformer Tap Switching (Fig. 11-22)

If a momentary interruption in power is acceptable, a time delay-on-operate is suggested to prevent overlap and the resulting high current surge from a shorted winding. Most AC input SSRs will already include enough delay to prevent make-before-break, otherwise a dual-on situation must be tolerated. Two times R_X plus the winding resistance must be sufficient to limit the surge current to the one cycle surge rating of the

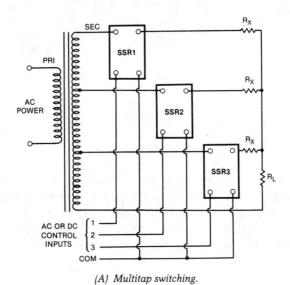

(A) Multitap switching.

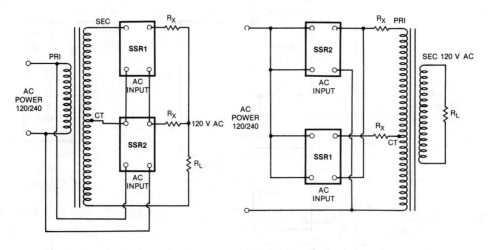

(B) Autotap selection (secondary).   *(C) Autotap selection (primary).*

Fig. 11-22. *Transformer tap switching.*

SSRs. As an additional precaution, the SSR blocking (breakdown) voltage should exceed the main winding voltage plus the highest tap voltage. For multitap switching (Fig. 11-22A) the SSRs are generally logic driven DC controlled AC output types without special requirements. For Figs. 11-22B and C, they are 240 V AC output with 120 V AC input for SSR1 and 240 V AC input for SSR2 in each case. An important requirement of SSR2 is that it must be off below the highest expected 120 V AC line swing, say 150 V AC. When SSR2 is off, SSR1 will be on and vice versa, thus activating the appropriate winding and preventing a possible catastrophic misapplication of voltage.

12

Testing the SSR

Many of the tests required to verify SSR performance are inherently hazardous and caution should be exercised, using adequate safeguards for the personnel conducting such tests. In field testing it should be noted that safety standards controlling power wiring, ducting, etc., are based on local electrical codes that can vary from one location to the next. However, industrial installations are generally constructed in compliance with the rules established in the National Electrical Code, ANSI/NFPA 70 and meet the applicable requirements of Underwriters Laboratories UL508 in the U.S.A.

Possibly the simplest of all field tests that can be made to determine proper function of an AC SSR is by means of a 3-volt battery, a light bulb, and a line cord (Fig. 12-1). Many field sales engineers carry such an apparatus with them to make a quick assessment of SSR performance in an unknown situation. For precise measurements of specific parameters, such as those made by receiving inspectors to verify specification limits, obviously more sophisticated test equipment is needed; however, the battery-bulb arrangement is useful for a quick failure analysis.

A more complete performance check might include operating the SSR in position with its actual load, while exercising the system installation functionally through all of its specified environmental and power combinations. Where this is not possible and SSR specification limits need to be checked, a general performance test, such as that defined in Section 12.9, may be more appropriate.

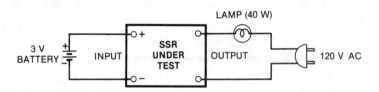

Fig. 12-1. *Simple go/no-go SSR test.*

When connecting test equipment directly to the power circuit of an SSR output, protective fusing would be a wise precaution. Also, remember that with some equipment such as an oscilloscope, the case must be "floated" (ungrounded) and may be at line potential. In some test circuits an isolated current probe or an isolation transformer can be used to avoid this hazard. The output functions of an SSR should not be checked (without power) by a VOM in the ohms mode, since the minimum voltages and bias currents necessary for proper SSR operation are not present, thus producing erroneous readings. (The battery voltage of the VOM is insufficient to produce semiconductor action on most ohms scales.)

The test methods in the following pages are set up for typical DC controlled AC SSRs using bidirectional thyristor outputs. DC SSRs, or those with AC inputs, require different treatment; however, the same general procedures may be followed.

An effort has been made to utilize only standard commonly available test equipment, thus avoiding the need for assembling special test circuits. The accuracy of standard laboratory equipment is generally more than adequate for most SSR testing. Precision equipment may occasionally be required to meet stringent user specifications. Test conditions are considered to be at normal room temperature and environment, close to 25°C (77°F).

For special investigative testing of parameters not defined in SSR or user specifications, such as endurance, life, temperature, moisture resistance, noise sensitivity, transient capability, EMI emissions and susceptibility, refer to test specifications listed in Appendix A.

12.1 Dielectric Strength (Isolation)

Dielectric strength is the most commonly made test to ensure isolation integrity. The test voltage (typically 2500 volts RMS, 60 hertz) is applied between input and output, and also input or output to case when the case has conductive (metallic) parts. The test duration is generally specified as one minute, (one second if applied voltage is 20 percent higher than that specified) with a maximum allowable leakage current of 1 milliampere.

(A) Test Setup: Fig. 12-2
(B) Equipment: AC hipot tester, 0-5 kilovolts, with built in ammeter for leakage measurement
(C) Procedure:
  1. Apply specified voltage between shorted input terminals and shorted output terminals.
  2. Observe relay under test and voltmeter for indication of breakdown for specified period. Also observe ammeter for allowable leakage.
  3. Apply specified voltage between all terminals shorted together and metal case and repeat 2, if applicable.

NOTE: If an SSR passes the dielectric test, manufacturer's stated values for insulation resistance and capacitance between the same electrodes are generally considered acceptable without further testing.

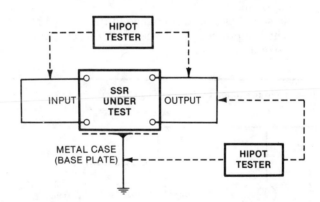

Fig. 12-2. *Dielectric (isolation) test*

12.2 Turn-On Voltage, Turn-Off Voltage, and Input Current

These parameters define the SSR input drive requirements in terms of must operate, must release, and input power. The turn-on voltage is typically equal to or greater than 3 volts DC and the turn-off voltage is typically equal to or less than 1 volt DC, with the band between these two values as an indeterminate state. For a normally closed SSR the on-off conditions would be reversed.

Often provided in SSR data sheets is an input control voltage range, the bottom end of which is usually the same as the turn-on voltage while the top end is a given maximum value limited by power dissipation in the input circuit.

The maximum current drain on the driving source is usually specified at two points—a low point and a high point within the control voltage range. This defines the input power requirements which may also be given in terms of input impedance at a given voltage.

(A) Test Setup: Fig. 12-3
(B) Equipment: Variable DC supply
                   Voltmeter *
                   Ammeter *
                   Lamp 120 volts, 40 watts
(C) Procedure:
  1. Slowly raise the DC control voltage until lamp lights. Observe turn-on voltage on voltmeter.
  2. Slowly lower the DC control voltage until lamp is extinguished. Observe turn-off voltage on voltmeter.
  3. Raise the DC control voltage to the level(s) at which input current is specified. Observe input current on ammeter.

* Instrument resolution to suit specification requirement.

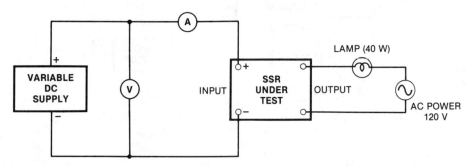

Fig. 12-3. *Test setup for measuring input turn-on, turn-off voltage, and input current.*

12.3 Turn-On and Turn-Off Times

These are defined as the maximum time between the application or the removal of the specified control signal and the transition of the output device to its appropriate fully conducting (on) or blocking (off) state. Response times usually include both propagation and rise or fall transition times, unless otherwise stated.

> (A) Test Setup: Fig. 12-4
> (B) Equipment: Variable DC supply
> Oscilloscope
> Variac (3 amperes)
> Mercury switch
> Load resistor (1000 ohms, 20 watts
> typical for 120 volts AC)
> (C) Procedure:
> 1. Set DC supply to suitable operating voltage
>   (greater than specified turn-on voltage).
> 2. Set variac to nominal output voltage.
> 3. Set scope to positive trigger.
> 4. Observe turn-on time on scope (Fig. 12-4B).
> 5. Set scope to negative trigger.
> 6. Observe turn-off time on scope (Fig. 12-4C).
> 7. Test may be repeated at high and low line voltage,
>   if required.

NOTE: A safer alternative test setup may be made with a dual beam
scope and a current probe to isolate the (hot) output signal.

12.4 Off-State Leakage

Defined as the RMS current that flows in the SSR output circuit while
it is in the offstate (no control signal applied). About half of the leakage
current results from drive circuit bias and leakage in the output power
device, and the other half is reactive (AC) leakage from the snubber
capacitor. Usually specified at nominal line voltage over the temperature
range.

> (A) Test Setup: Fig. 12-5
> (B) Equipment: Variable AC (60 hertz) supply with
> current limiting
> Voltmeter
> Milliammeter
> (C) Procedure:
> 1. Set current limiter to value that will not interfere
>   with peak leakage reading, but will protect
>   milliammeter should SSR turn on.
> 2. Raise voltage to value at which leakage current is
>   specified.
> 3. Observe leakage reading on milliammeter.

NOTE: When required, procedure may be repeated with test relay in
temperature chamber at temperature limits.

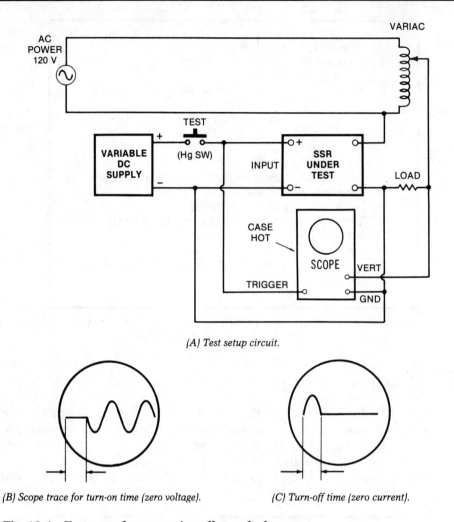

(A) Test setup circuit.

(B) Scope trace for turn-on time (zero voltage).   *(C) Turn-off time (zero current).*

Fig. 12-4. Test setup for measuring off-state leakage.

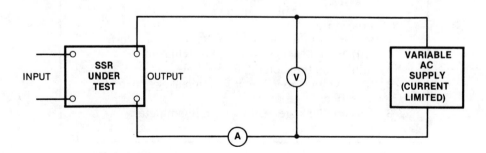

Fig. 12-5. Test setup for measuring off-state leakage.

12.5 Transient Overvoltage (Nonrepetitive Peak Voltage)

Transient overvoltage is commonly known as blocking voltage and is defined as the maximum allowable peak off-state voltage that an SSR can withstand without breakdown. This value is dissipation limited by the SSR circuitry, possibly causing damage if sustained; however, it may be held long enough to take a reading during the following test.

(A) Test Setup: Fig. 12-6
(B) Equipment: Oscilloscope
Variac (to suit specified SSR voltage)
Load resistor (1000 ohms, 20 watts typical for 120 volts AC)
(C) Procedure:
1. Set variac to nominal voltage and adjust scope.
2. Momentarily raise variac voltage to the peak value specified for SSR.
3. Observe oscilloscope for any indication of breakdown.

NOTE: Do not leave voltage at high setting for more than 30 seconds.

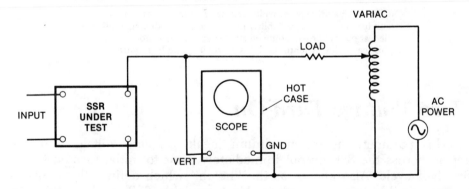

Fig. 12-6. *Test setup for checking transient overvoltage (nonrepetitive peak voltage).*

12.6 Peak Surge Current (Nonrepetitive)

This is the maximum allowable momentary output current flow, typically specified as a peak value for one line cycle (16.6 milliseconds). If not specified as a separate parameter, it may be taken from time versus current curves in manufacturers' data sheets. In the following procedure the test waveshape is sinusoidal.

(A) Test Setup: Fig. 12-7
(B) Equipment: Variable DC supply
Variable pulse generator
Time delay (10 millisecond)
Current-limiting resistors
SSR 240 volt, 75-90 ampere (D2475)
zero switching type
Variac (100 ampere)
Welding transformer (1000 ampere)
Oscilloscope
Monitor shunt (50 millivolt)
Switch
(C) Procedure:
1. Mount SSR to suitable heat sink if test repetition rate is less than one minute.
2. Set DC supply to suitable operating voltage (greater than specified turn-on voltage).
3. Set current-limiting resistor to appropriate range.
4. Set pulse generator to 15 milliseconds.*
5. Observe applied current pulse on oscilloscope and increase variac voltage until desired peak value is reached.
6. Repeat test at extremes of SSR control voltage, if required.

*NOTE: With a drive pulse width setting of 15 milliseconds this circuit has the highest probability (90 percent) of producing a full cycle test pulse. For a guaranteed single cycle pulse a more sophisticated line synchronized drive circuit would be required.

12.7 Zero Voltage Turn-On

This parameter is the maximum (peak) off-state voltage that can appear across the SSR output immediately prior to initial turn-on in the first half cycle. Also known as the "notch" which defines the limits of the permissible turn-on window. Most available SSRs include this feature as opposed to random turn-on types, which are generally used for phase control.

(A) Test Setup: Fig. 12-8
(B) Equipment: Variable pulse generator with
                        delayed trigger
                        Oscilloscope
                        Filament transformer (6.3 volt)
                        Nominal load resistor
(C) Procedure:
  1. Set pulse amplitude equal to or greater than SSR
      turn-on voltage.
  2. Set pulse width to approximately 2 milliseconds
      and set for positive trigger.
  3. Adjust trigger pulse delay until SSR turn-on is
      observed on oscilloscope.
  4. Slowly increase delay observing increase in turn-
      on voltage until SSR fails to turn on. Maximum
      zero turn-on voltage is the highest peak seen
      before failure to turn on (trailing edge of turn-on
      pulse).

NOTE: Turn-on pulse may be observed simultaneously by using a
dual beam oscilloscope and an isolation transformer. Also, a simpler
alternative test may be implemented on a curve tracer, if available.

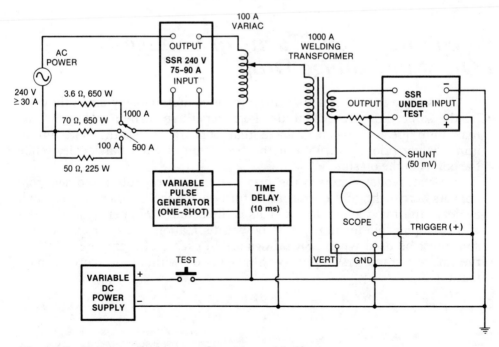

Fig. 12-7. *Test setup for surge and overcurrent testing.*

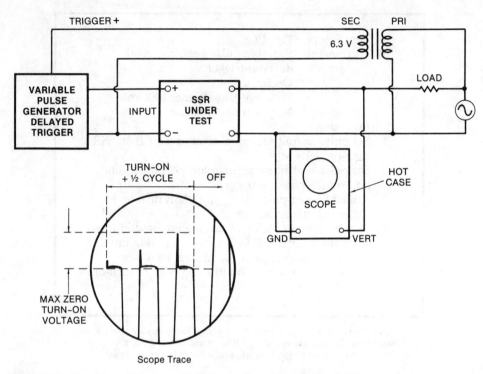

Fig. 12-8. *Test setup for measuring zero voltage turn-on (notch).*

12.8 Peak Repetitive Turn-On Voltage and On-State Voltage Drop

These two measurements can be made simultaneously since they are on the same order of magnitude. Peak repetitive turn-on voltage is the maximum (peak) off-state voltage that appears across the SSR output immediately prior to turn-on at the beginning of each half cycle, while the SSR is in the on state.

On-state voltage drop is the maximum (peak repetitive) voltage that appears across the SSR output after turn-on for the balance of each half cycle, at full rated load current. This is the actual "on" period of the SSR and is responsible for most of its power dissipation. An appropriate heat sink must be used when SSR ratings are in excess of 3 amperes. Both of the above parameters apply to SSRs with or without the zero turn-on feature.

(A) Test Setup: Fig. 12-9
(B) Equipment: DC control source (equal to or
greater than specified turn-on
voltage)
Oscilloscope
Specified load
Heat sink (if required)
(C) Procedure:
1. Apply turn-on control voltage.
2. Set oscilloscope horizontal for two or three cycles
(line synchronized) and vertical to about 2 volts
per centimeter.
3. Observe peak repetitive on voltage and on-state
voltage drop as shown on scope trace (Fig. 12-9).

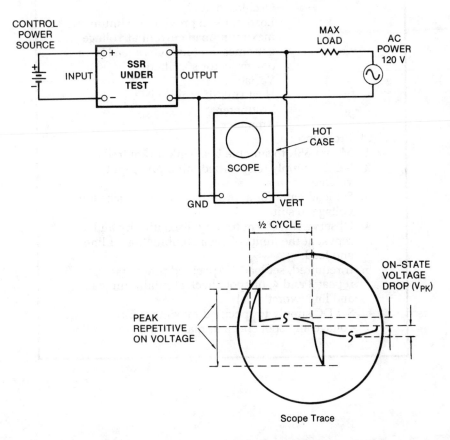

Fig. 12-9. *Test setup for measuring peak repetitive turn-on voltage and on-state voltage drop.*

12.9 Operating Voltage and Load Current Ranges

This test is considered to be the most comprehensive performance bench test, where the SSR is taken to the extremes of its steady-state output capability. The output waveform is generally viewed across the load to verify a reasonably distortion-free replica of the applied line voltage, under minimum and maximum load current conditions including inductive loads where applicable.

For loads typically in excess of 3 amperes, a heat sink must be selected that will permit continuous SSR operation at room ambient temperature.

(A) Test Setup: Fig. 12-10

(B) Equipment: Variable DC supply
Oscilloscope
Load bank to provide minimum and maximum load current at voltage extremes, and inductive load with power factor specified
Variac
Heat sink (if required)
Voltmeter
Ammeter

(C) Procedure:

1. Mount SSR to suitable heat sink if required.
2. Set DC supply to specified minimum input voltage.
3. Set load bank for appropriate load and raise line voltage to suit.
4. Observe waveform for any discontinuity and repeat at the limits of each combination of line and load.
5. If required, set load to specified power factor and repeat 3 and 4, or spot check at minimum load and line (worst case).
6. Set DC supply to specified maximum input voltage and spot check at maximum load and line.

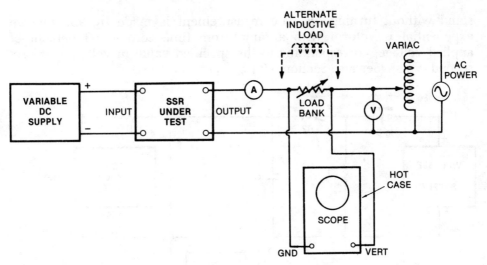

Fig. 12-10. *Test setup for operating voltage and load current ranges.*

12.10 dv/dt (Rate Effect)

The given parameter, static (off-state) dv/dt, is defined as the rate of rise of applied voltage across the output terminals that an SSR can with

---

(A) Test Setup: Fig. 12-11
(B) Equipment: Variable DC supply (low ripple)
Mercury switch
Bleed resistor
Variable resistor or carbon pile
Supply capacitor
Source/limit resistor (typical 50 ohm)
Oscilloscope
(C) Procedure:
  1. Raise DC level to specified voltage.
  2. With SSR connected and R2 at maximum (slow ramp), periodically operate test switch while proportioning waveshape (positive trigger).
  3. While periodically operating test switch reduce R2 (increasing ramp) until specified value is reached without mistrigger.
  4. Reverse SSR connections and with same settings, operate test switch while observing oscilloscope for failure.

NOTE: An SSR failure (mistrigger) is observed as a sudden drop in voltage interrupting the exponential rise. Settings may remain unchanged when testing SSRs of the same type.

stand without turning on. The measurement is made by applying an exponential waveform with a ramp from time zero to 63 percent of applied voltage, corresponding to the specified value in volts per microsecond (V/μs). (See also Section 9.2.)

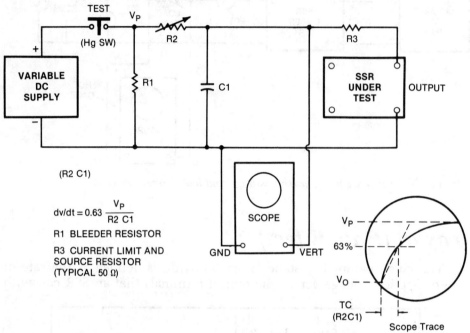

$$(R2\ C1)$$

$$dv/dt = 0.63 \, \frac{V_P}{R2 \ C1}$$

R1 BLEEDER RESISTOR

R3 CURRENT LIMIT AND
SOURCE RESISTOR
(TYPICAL 50 Ω)

Fig. 12-11. *Test setup for dv/dt test.*

12.11 *Failure Modes*

Systems reliability analysts sometimes need to know the probability of failure of an SSR in a particular mode, under certain conditions, in order to predict system consequences. Assuming the SSR incorporates commonly used discrete circuitry, a random internal component failure, open or closed, will yield a greater than 50 percent probability of the output failing in the open mode. Failure due to a recurring component defect will obviously slant the failure mode in a predictable direction until the problem is eliminated.

Based on records of SSR field returns where abuse was the detectable reason for failure (i.e., overvoltage, current, or temperature), failure in the closed (on) state is a 90 percent probability. The 10 percent open failure mode is most frequently due to a secondary effect where, for example, a shorted load results in a shorted SSR output, which in turn burns open the internal current path due to inadequate fusing. The latter will usually show evidence of overheating or burning.

Lifetime predictions are rarely made by commercial SSR manufacturers and when made, they are based on the military "MIL HDBK 217" method. Predictions made by this method provide a mean time between failure (MTBF) which is little more than a guess unless the manufacturer is using military grade components and processes. Occasionally, curves showing SSR lifetime versus current surge occurrences are given (Section 8.1), but these do not relate to actual steady-state lifetime.

Some manufacturers do provide a limited amount of lifetime data based on actual accumulated operating hours for a given number of SSRs. This type of data, usually developed for a special application or user specification, is considered the most valuable. In the absence of suitable reliability data, a user can take some comfort in the fact that a UL recognized SSR at one time had to withstand the overload and endurance testing of UL 508.

The erratic performance of an SSR may be caused by an intermittent internal component, or it may be application related. A quick determination as to which is at fault can usually be made by replacing the suspect SSR with a second one. If after replacement the problem remains, the application should be examined. Failures of a nonpermanent nature are generally application related and the solutions to such problems are covered in depth in Chapters 6 through 9.

13

Manufacturers and Cross-Reference Diagrams

As previously stated, during the first decade since SSRs were introduced, manufacturers went off in many different directions in both electrical and physical parameters. However, in recent years some standardization has occurred, even to the extent of commonality in part numbers, some of which are listed in the following cross-reference diagrams. While the cross-reference diagrams may assist in the selection of the popular varieties, they are but a small fraction of those presently available from the more than three dozen manufacturers listed in Section 13.2.

In selecting a suitable current rating, reference must be made to the manufacturer's derating curves where the usable value is related to ambient or case temperature. The chapter on thermal considerations (Chapter 7) would be a helpful aid in making this selection and that of an appropriate heat sink.

Caution should be exercised when reviewing manufacturers' specifications at face value. For example, vendor A may have a 4-ampere, 120-volt unit in a given package, while vendor B has a similar 3-ampere, 120-volt unit in the same package. Upon examination and comparison of their respective thermal curves, it may be found that the vendor A unit is indeed rated at 4 amperes but only at 20°C or less (below room temperature), while the vendor B unit is rated for 3 amperes at 40°C (Fig. 13-1). In this case the 3-ampere unit is actually 5 degrees or one third of an ampere better over the derated temperature range and would, in fact, be a more believable 4-ampere device at 25°C had the curve been extended.

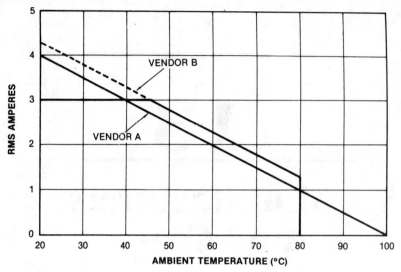

Fig. 13-1. *Derating curves.*

13.1 Available SSR Package Styles

The SSR package styles illustrated in pages 177–185 are identified by letter, the availability of which is shown parenthesized under the listing for each manufacturer. In addition to those shown, many packages will have other terminal options (e.g., screw terminals, pins or fast-ons).

COMPATIBLE CHASSIS MOUNTABLE SSRs WITH DUAL SCR OUTPUTS

Input Voltage	Output Voltage AC	Blocking Voltage (peak)	Output Amperes	Crydom	Opto-22	Gordos	Magnacraft
90-280(AC)	24-140	250	10	A1210	120A10	G120A10	W6110ASX-1
90-280(AC)	24-140	250	25	A1225	120A25	G120A25	W6125ASX-1
90-280(AC)	24-140	250	40	A1240	120A45	G120A45	W6140ASX-1
90-280(AC)	48-280	500	10	A2410	240A10	G280A10	W6210ASX-1
90-280(AC)	48-280	500	25	A2425	240A25	G280A25	W6225ASX-1
90-280(AC)	48-280	500	40	A2440	240A45	G280A45	W6240ASX-1
90-280(AC)	80-480	800	10	A4812	480A10-12	G480A10	W6412ASX-1
90-280(AC)	80-480	800	25	A4825			W6425ASX-1
90-280(AC)	80-480	800	40	A4840			
3-32 (DC)	24-140	250	10	D1210	120D10	G120D10	W6110DSX-1
3-32 (DC)	24-140	250	25	D1225	120D25	G120D25	W6125DSX-1
3-32 (DC)	24-140	250	40	D1240	120D45	G120D45	W6140DSX-1
3-32 (DC)	48-280	500	10	D2410	240D10	G280D10	W6210DSX-1
3-32 (DC)	48-280	500	25	D2425	240D25	G280D25	W6225DSX-1
3-32 (DC)	48-280	500	40	D2440	240D45	G280D45	W6240DSX-1
3-32 (DC)	80-480	800	10	D4812	480D10-12	G480D10	W6412DSX-1
3-32 (DC)	80-480	800	25	D4825			
3-32 (DC)	80-480	800	40	D4840			

(Note: Some manufacturers exceed these ratings)

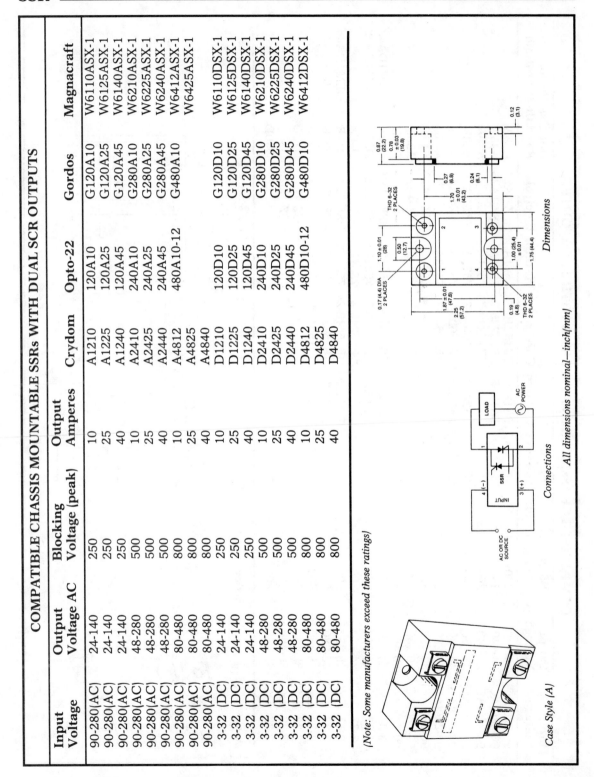

Dimensions

All dimensions nominal—inch(mm)

Connections

Case Style (A)

COMPATIBLE CHASSIS MOUNTABLE SSRs WITH TRIAC OUTPUT

Input Voltage	Output Voltage AC	Blocking Voltage (peak)	Output Amperes	Crydom	Gordos	Hamlin	Electrol	Douglas Randall	Potter & Brumfield
90-280(AC)	24-140	200	10	TA1210	GB15210-7	7521A	HSZ-1A10	A12A	OEM1DA42
90-280(AC)	24-140	200	25	TA1225	GB15225-7	7531A	HSZ-1A25	A25A	OEM1DA72
90-280(AC)	24-140	200	40		GB15240-7	7551A		A40A	
90-280(AC)	48-280	500	10	TA2410	GB15610-7	7523A	HSZ-2A10	A12B	OEM1DA44
90-280(AC)	48-280	500	25	TA2425	GB15625-7	7533A	HSZ-2A25	A25B	OEM1DA74
90-280(AC)	48-280	500	40		GB15640-7	7553A		A40B	
90-280(AC)	80-480	800	10				HSZ-4A10	A12D*	
90-280(AC)	80-480	800	25				HSZ-4A25	A25D*	
90-280(AC)	80-480	800	40					A40D*	
3-32 (DC)	24-140	200	10	TD1210	GB15210-2	7521D	HSK-1D10	D12A	
3-32 (DC)	24-140	200	25	TD1225	GB15225-2	7531D	HSK-1D25	D25A	
3-32 (DC)	24-140	200	40		GB15240-2	7551D		D40A	
3-32 (DC)	48-280	500	10	TD2410	GB15610-2	7523D	HSK-2D10	D12B	
3-32 (DC)	48-280	500	25	TD2425	GB15625-2	7533D	HSK-2D25	D25B	
3-32 (DC)	48-280	500	40		GB15640-2	7553D		D40B	
3-32 (DC)	80-480	800	10				HSZ-4D10	D12D*	
3-32 (DC)	80-480	800	25				HSZ-4D25	D25D*	
3-32 (DC)	80-480	800	40					D40D*	

(Note: Some manufacturers exceed these ratings)

*No Snubber

Connections

Dimensions

All dimensions nominal—inch(mm)

Case Style (A)

MEDIUM POWER SSRs WITH FAST-ON TERMINALS AND TRIAC OUTPUT

Input Voltage DC	Output Voltage AC	Blocking Voltage (peak)	Output Amperes	Antex/ Guardian	Potter & Brumfield	Grayhill	Opto-22	Sigma	Midtex
4-6	75-140	200	10	SSR-1B-10	EOT1DB425	70S2-03 -B-10-D	Z120D10	223A-3-5D	640-13L200
4-6	100-240	400	10	SSR-2B-10	EOT1DB445	70S2-03 -C-10-D	Z240D10	—	640-23L200
4-6	75-140	200	4 (7)	SSR-1A-10	EOT1DB425	70S2-03 -B-06-D	Z120D5	223A-1-5D	640-11L200
4-6	100-240	400	4 (7)	SSR-2A-10	EOT1DB445	70S2-03 -C-06-D	Z240D5	—	640-21L200

Note: 1. Some manufacturers have wider input ranges, but all may be switched by 5 volts (TTL).
   2. 10 ampere models must be mounted on a chassis (heat sink).
   3. All zero voltage turn-on except 223A/-3-5D/-1-5D.

Dimensions

Connections

Case Style (B)

PLUG-IN I/O MODULES (TRIAC AND TRANSISTOR OUTPUTS)

Input Voltage	Logic Supply Voltage DC	Output Voltage	Output Current	Crydom	Gordos	Opto-22	Potter & Brumfield	Electrol
10-32 (DC)	4.5-6	30(DC)	25 MA	6101A	IDC5	IDC5	IDC5	IDC5
10-32 (DC)	12-18	30(DC)	25 MA	6131	IDC15	IDC15	IDC15	IDC15
10-32 (DC)	20-30	30(DC)	25 MA		IDC24	IDC24	IDC24	IDC24
90-140(AC)	4.5-6	30(DC)	25 MA	6201A	IAC5	IAC5	IAC5	IAC5
90-140(AC)	12-18	30(DC)	25 MA	6231	IAC15	IAC15	IAC15	IAC15
90-140(AC)	20-30	30(DC)	25 MA		IAC24	IAC24	IAC24	IAC24
180-280(AC)	4.5-6	30(DC)	25 MA	6202A	IAC5-A	IAC5-A	IAC5-A	IAC5-A
180-280(AC)	12-18	30(DC)	25 MA	6232	IAC15-A	IAC15-A	IAC15-A	IAC15-A
180-280(AC)	20-30	30(DC)	25 MA		IAC24-A	IAC24-A	IAC24-A	IAC24-A
3-6 (DC)		60(DC)	3 A	6301A	ODC5	ODC5	ODC5	ODC5
9-18 (DC)	(*10-18)	60(DC)	3 A	*6341	ODC15	ODC15	ODC15	ODC15
20-28 (DC)		60(DC)	3 A		ODC24	ODC24	ODC24	ODC24
3-6 (DC)		12-140(DC)	3 A	6401A	OAC5	OAC5	OAC5	OAC5
9-18 (DC)	(*10-18)	12-140(AC)	3 A	*6441	OAC15	OAC15	OAC15	OAC15
20-28 (DC)		12-140(AC)	3 A		OAC24	OAC24	OAC24	OAC24
3-6 (DC)		24-280(AC)	3 A	6402A	OAC5-A	OAC5-A	OAC5-A	OAC5-A
9-18 (DC)	(*10-18)	24-280(AC)	3 A	*6442	OAC15-A	OAC15-A	OAC15-A	OAC15-A
20-28 (DC)		24-280(AC)	3 A		OAC24-A	OAC24-A	OAC24-A	OAC24-A

*These models have low power buffered inputs and require a logic supply.

Dimensions

Connections

Case Style (C)

SMALL PLUG-IN SSRs WITH TRIAC AND TRANSISTOR OUTPUTS

Input Voltage	Output Voltage	Blocking Voltage (peak)	Output Amperes	Crydom	Opto-22	Hamlin	Gordos	Guardian
3.5-15(DC)	12-140(AC)	200	2		MP120D2		GA8-2B01	ISSR-1H-10
3.5-15(DC)	12-140(AC)	200	3	MP120D3	MP120D3	7591	GA8-2B02	ISSR-1H-30
3.5-15(DC)	24-280(AC)	400	2		MP240D2	7592	GA8-4B01	ISSR-2H-10
3.5-15(DC)	24-280(AC)	400	3	MP240D3	MP240D3	7593	GA8-4B02	ISSR-2H-30
3.5-18(DC)	4-60 (DC)	60	3	MP-DCD3	DC60MP			

Notes: 1. Input voltage ranges vary slightly.
2. Some manufacturers exceed these ratings.

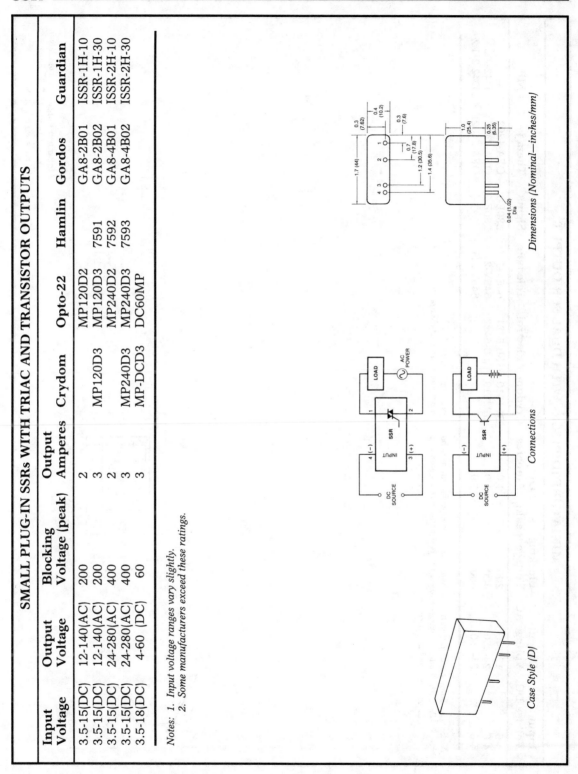

Dimensions (Nominal—inches/mm)

Connections

Case Style (D)

MINIATURE (DIP) SSRs WITH DUAL SCR OUTPUT

Input Voltage DC	Output Voltage AC	Blocking Voltage (peak)	Output Amperes	Crydom	Electrol	Teledyne	Sigma	Theta J	MR1
4-6	12-140	200	1.0	DP1610	SB-5421	645V-1	230E-1	OFA-1202	—
4-6	24-250	400	1.0	DP2610	SB-5422	645V-2	230E-2	OFA-2402	D42-2-2
4-6	24-250	600	1.0			645V-2H		OFA-2402H	D42-2H-2

Notes: 1. Some manufacturers have wider input ranges but all may be switched by 5 volts (TTL).
2. Output current varies somewhat, generally in the region of one ampere without heat dissipator.
3. All zero voltage turn-on.

Connections

Dimensions

Max dimensions—some models in standard 16 pin DIP package. All have same pin-out.

(E)

(F)

Case Styles

MISCELLANEOUS CASE STYLES

Case Style (G)

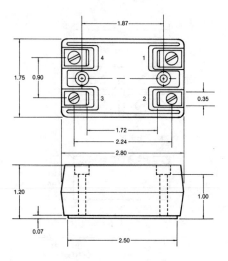

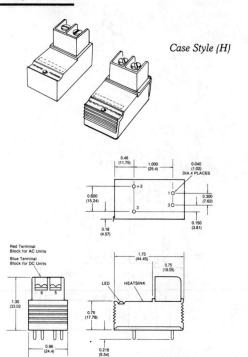

Case Style (H)

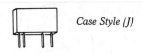

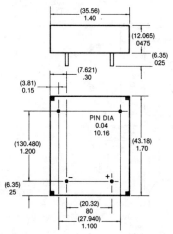

Case Style (I)

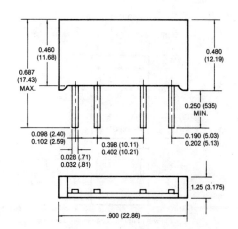

Case Style (J)

MISCELLANEOUS CASE STYLES

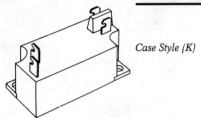

Case Style (K)

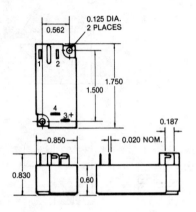

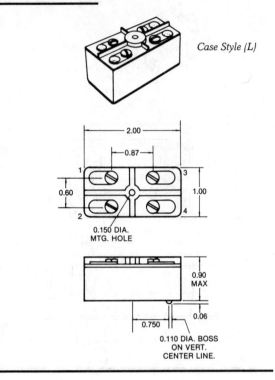

Case Style (L)

0.150 DIA.
MTG. HOLE

0.90 MAX

0.06

0.750

0.110 DIA. BOSS
ON VERT.
CENTER LINE.

Case Style (M)

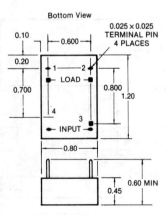

Bottom View

0.025 x 0.025
TERMINAL PIN
4 PLACES

0.10
0.20
0.700
0.600
1 — 2
LOAD
0.800
1.20
4 — 3
INPUT
0.80
0.60 MIN
0.45

Pin will plug into a 24-pin DIP socket

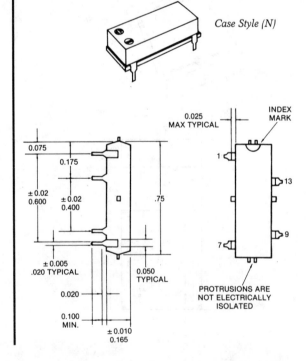

Case Style (N)

0.025 MAX TYPICAL

INDEX MARK

PROTRUSIONS ARE
NOT ELECTRICALLY
ISOLATED

0.075
0.175
± 0.02
0.600
± 0.02
0.400
.75
± 0.005
.020 TYPICAL
0.050 TYPICAL
0.020
0.100 MIN.
± 0.010
0.165

MISCELLANEOUS CASE STYLES

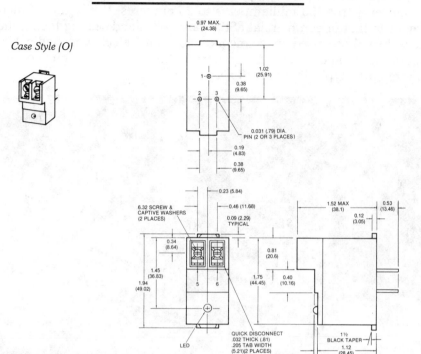

Case Style (O)

13.2 Directory of Manufacturers and Suppliers of SSRs

Parenthesized letters indicate package styles available from those listed in previous pages. (SP) indicates special packages not listed.

While the packages shown on pages 177–185 are the more popular current types, others will undoubtedly emerge from advancing technologies to replace many of them. Present indications are that most of the newer designs are headed towards miniaturization by utilizing monolithic and hybrid techniques in their construction. The power types, however, will shrink very little in the mounting base area due to the necessity of heat transfer.

The increasing use of robotic assembly techniques is also bound to influence SSR design. The economy, speed, and reliability are the main attractions. Shown in Fig. 13-2 is the center portion of an automatic assembly machine for production of SSR surface mount PC boards. Note circuit boards proceeding by assembly stations where individual robotic heads place components which are organized by the associated feeder bowls and cartridges.

The SSR packages on pages 177–185 are all of the industrial-commercial variety which is their primary market. A smaller select group of SSRs exists that was developed specifically for the military market. Some of these, shown in Fig. 13-3, are hermetically sealed ranging in load switching currents from 50 milliamperes to 25 amperes. While their function is similar to that of commercial SSRs, case styles and sealing techniques are more like those used in semiconductors and electromechanical relays. Operating temperatures are generally higher than their commercial counterparts, typically up to 125°C.

Fig. 13-2. *The robotic assembly of SSR surface mount PC boards.* *(Courtesy IR Crydom)*

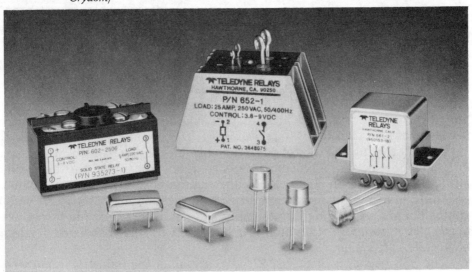

Fig. 13-3. *A sampling of military style SSR packages.* *(Courtesy Teledyne Solid State Products)*

ANTEX ELECTRONICS CORP.
16100 S. Figueroa St.
Gardena, CA 90248
TEL: (213) 532-3092
TWX: 910-344-7381

(A) (B) (C) (D) (I)

APPLIED ELECTROTECHNOLOGY, INT.
2220 S. Anne St.
Santa Ana, CA 92704
TEL: 714-556-6570
TWX: 910-595-1797

(A) (I)

CAPITAL POWER DEVICES LTD.
Westbourne St.
High Wycombe
England
TEL: (0494) 38247
TLX: 933469

(A) (B) (D)

CELDUC
57, Rue De Prony
75017 Paris
France
TEL: 267.14.10
TLX: 640256

(A) (C) (D) (M)

CONTINENTAL INDUSTRIES, INC.
5456 E. McDowell Rd.
Mesa, Arizona 85205
TEL: (602) 985-7800

(A) (B) (C) (D)

CRYDOM DIVISION INTERNATIONAL RECTIFIER
1521 East Grand Avenue
El Segundo, CA 90245
TEL: (213) 322-4987
TWX: 910-348-6283

(A) (C) (D) (E) (H) (J) (K) (M)

DIONICS, INC.
65 Rushmore St.
Westbury, NY 11590
TEL: (516) 977-7474
TWX: 510-222-0974

(SP)

DOUGLAS RANDALL DIV.
Walter Kidde & Co.
6 Pawcatuck, CT 02891
TEL: (203) 599-1750

(A)

ELEC-TROL, INC.
26477 North Golden Valley Rd.
Saugus, CA 91350
TEL: (805) 252-8330
TLX: 181151

(A) (C) (E)

ELECTROMATIC
DK-8370
Hadsten
Denmark
TEL: 456 981100
TLX: 60786

(A) (D)

**ELECTRONIC INSTRUMENT &
SPECIALTY CORPORATION**
42 Pleasant St.
Stoneham, MA 02180
TEL: (617) 729-8700
     (617) 438-5300

(A) (D) (F) (M)

FR ELECTRONICS
7 Cobham Rd.
Ferndown Ind. Est.
Wimbourne, Dorset BH217PE
England
TEL: 0202-897969
TLX: 41391

(A) (C) (D) (I)

**GORDOS INTERNATIONAL
CORP.**
1000 N. 2nd St.
Rogers, Arkansas 72756
TEL: (501) 63605000
TWX: 910-720-7998

(A) (B) (C) (D) (I) (M)

**GOULD, INC.
INDUSTRIAL CONTROLS
DIVISION**
100 Relay Road
Plantsville, CT 06479
TEL: (203) 621-6771
TWX: 710-473-0079

(A)

GRAYHILL, INC.
561 Hillgrove Ave.
La Grange, IL 60525-0373
TEL: (312) 354-1040
TWX: 910-683-1850

(B) (L)

**GUARDIAN ELECTRIC
MANUFACTURING COMPANY**
1550 W. Carrol Ave.
Chicago, IL 60607
TEL: (312) 243-1100
TWX: 910-221-5073

(A) (B) (C) (D) (I)

HAMLIN, INC.
612 East Lake St.
Lake Mills, WI 53551
TEL: (414) 648-2361
TWX: 910-260-3740

(A) (B) (D)

HASCO COMPONENTS, INC.
214-22 Waters Edge Dr.
Bayside, NY 11360
TEL: (212) 428-5155
TWX: 710-582-5750

(SP)

HEINEMANN ELECTRIC CO.
P.O. Box 6800
Lawrenceville, NJ 08648-0800
TEL: (609) 882-4800
TLX: 843431

(A) (L)

IDEC SYST. & CONTR. CORP.
3050 Tasman Dr.
Santa Clara, CA 95050
TEL: (408) 988-7500
TLX: 340601

(A)

ITT COMPONENTS
1201 East McFadden Ave.
Santa Ana, CA 92705
TEL: (714) 836-0351
TWX: 910-595-1744

(A) (B) (C)

KIPCO CORP.
P.O. Box 2578
Syracuse, NY 13220
TEL: (315) 699-5266
TWX: 710-545-0235

(SP)

**LEACH CORPORATION
RELAY DIVISION**
6900 Orangethorpe Ave.
Buena Park, CA 90620
TEL: (714) 739-1150

(SP)

**MAGNECRAFT ELECTRIC
COMPANY**
5575 North Lynch Avenue
Chicago, IL 60630
TEL: (312) 282-5500
TWX: 910-221-5221

(A) (M)

**MASTER ELECTRONIC
CONTROLS**
P.O. Box 25905
Los Angeles, CA 90025
TEL: (213) 452-1336
TLX: 182041

(A) (H)

**MICROELECTRONIC RELAYS,
INC.**
1300 S. Beacon St.
San Pedro, CA 90731
TEL: (213) 831-2331

(A) (E) (J) (N)

**MIDLAND ROSS CORP.
MIDTEX DIVISION**
1650 Tower Blvd.
North Mankato, MN 56001
TEL: (507) 625-6521
TWX: 910-565-2244

(B)

MSI ELECTRONICS, INC.
34-32 57th St.
Woodside, NY 11377
TEL: (212) 672-6500

(A) (D) (I) (M)

OMEGA ENGINEERING, INC.
1 Omega Dr.
Box 4047, Stamford, CT 06907
TEL: (203) 359-1660
TLX: 996404

(A) (C)

OMRON ELECTRONICS, INC.
1 East Commerce Dr.
Schaumburg, IL 60195

(SP)

OPTO-22
15461 Springdale St.
Huntington Beach, CA 92649
TEL: (714) 891-5861
TLX: 692386

(A) (B) (C) (D) (I)

**PHILIPS ECG, INC.
DEPT. EE**
1025 Westminster Dr.
Williamsport, PA 17701

(A)

**POTTER & BRUMFIELD
DIVISION OF AMF, INC.**
200 Richland Creek Drive
Princeton, IN 47671
TEL: (812) 386-1000

(A) (B) (C) (I)

SIEMENS CORPORATION
Electromagnetic Relays
240 East Palais Rd.
Anaheim, CA 92805

(A) (D)

SIGMA INSTRUMENTS, INC.
170 Pearl St.
Braintree, MA 02184
TEL: (617) 843-5000
TLX: 940645

(B) (F)

SILICON POWER CUBE
2725 Seaboard Ln.
P.O. Box 5516
Long Beach, CA 90805
TEL: (213) 634-9390

(A)

STRUTHERS-DUNN, INC.
Lambs Road
Pitman, NJ 08071
TEL: (609) 589-7500
TWX: 510-686-7510

(G) (N)

TECCOR
P.O. Box 61447
Dallas, TX 75261
TEL: (214) 252-7651

(A) (B)

**TELEDYNE SOLID STATE
PRODUCTS**
12525 Daphne
Hawthorne, CA 90250
TEL: (213) 777-0077
TWX: 910-321-4610

(A) (E) (G) (L) (N) (O)

THETA-J
8 Corporate Pl.
107 Audubon Rd.
Wakefield, MA 01880
TEL: (617) 246-4000
TWX: 310-681-7300

(E) (SP)

**TOSHIBA AMERICA, INC.
ELECTRONIC COMPONENTS
DIVISION**
2441 Michelle Dr.
Tustin, CA 92680
TEL: (714) 730-5000
TLX: 182312

(A) (D)

14

Glossary of Commonly Used SSR Industry Terms

AC Alternating current. Also used to designate a sinusoidal voltage that causes a current of alternating polarity to flow in a resistive load.

Allowable Input Current (Maximum) Tolerable leakage current from a control source that will not cause a change in the SSR output state (generally while "off").

Ambient Temperature The surrounding air temperature usually specified with upper and lower limits for both operating and storage.

Ampere Unit of measure of electrical current. One ampere is the current which will flow through a one-ohm resistor when an electromotive force of one volt is applied.

Anode High potential terminal of an SCR. Positive in respect to gate and cathode when conducting (blocking when negative).

Armature The moving magnetic member of an EMR.

Base The control terminal of a bipolar transistor.

Bidirectional Essentially the same switching behavior and current conducting capability in both directions (positive or negative).

Bipolar Generally used to describe a transistor type in which a DC current flow between collector and emitter is modulated by a smaller current flowing between base and emitter. The gain of the transistor

191

relates to the ratio of these two currents defined as beta or h_{FE} in common-emitter configurations.

Bistable   A two-state device that will remain in its last operated state after control power is removed (e.g., latching relay).

Blocking Voltage   Maximum allowable standoff voltage before breakdown.

Blocking (Reverse)   Maximum allowable reverse standoff voltage for SCR with negative anode to cathode voltage.

Bounce, Contact   Intermittent and undesired opening of closed contacts or closing of open contacts of an EMR.

Break   The opening or interruption of an electric current.

Breakdown (Breakover)   The point at which blocking capability in an SSR collapses as voltage is increased beyond its maximum (transient) rating.

Bridge   A special connection of rectifier diodes used in the conversion of AC to DC.

Capacitance   The ability to store an electrical charge. Also given as an SSR isolation parameter, measured input to output, or both to case, provided as a means of determining high-frequency noise coupling.

Case   The housing or base plate of an SSR. A point on the case is sometimes defined for temperature measurement.

Cathode   SCR terminal associated with gate terminal. Negative in respect to anode when conducting.

Clapper Relay   Sometimes used to describe an EMR.

CMOS   Complementary metal oxide semiconductor. An IC logic family using complementary P and N channel MOSFET devices in a configuration that requires very little standby current.

Coil   A winding usually wound over an insulated iron core which, when energized, will operate the armature of an EMR.

Collector   A main current terminal and also high-voltage terminal of a transistor relative to the base and emitter.

Common-Mode Noise   The application of high-frequency noise between input and output.

Commutation   The transfer of current flow from one circuit element to another.

Conductor   A material which allows easy flow of current.

Contactor   Power EMR (usually greater than 10 amperes).

Contacts The current-carrying members of an EMR that open or close electric circuits. Sometimes used to denote the output of an SSR.

Control Voltage Specified as a range of voltages which, when applied across the SSR input terminals, will maintain an on condition across the output terminals (normally open).

Crosstalk The electrical coupling between two electrical circuits.

Current The rate of flow of electricity. (See ampere.)

Cycle A complete sequence of events; generally repeated in cycles per second, as in the repetition rate of alternating wave. (See hertz.)

DC (Direct Current) Continuous current or voltage of a given amplitude and polarity.

di/dt Maximum rate of rise of on-state load current that an SSR can withstand without damage. A characteristic of thyristors used in AC SSRs.

dv/dt Maximum rate of rise of voltage applied across the output terminals that the SSR can withstand without turning on. A characteristic of thyristors used in AC SSRs.

dv/dt–Off State (Static) Specified as a minimum dv/dt withstand capability of SSRs in the off or blocking state.

dv/dt–Commutating A thyristor parameter specified as the maximum rate of rise of reapplied voltage immediately following conduction that will allow a triac to regain its blocking capability (at turn-off).

Darlington High gain combination of two transistors cascaded to compound their respective gains.

Deenergize The removal of power from a relay coil or control input.

Dielectric Strength The maximum allowable AC RMS voltage (50/60 hertz) which may be applied between two specified test points.

Diode A semiconductor which allows electric current to flow easily in one direction but not in the other. (See rectifier.)

Discrete Device An individual electrical component such as a resistor, capacitor, or transistor, as opposed to an IC that is equivalent to several discrete components.

Dropout An EMR drops out when it changes from an energized condition to the unenergized condition. (See turn-off voltage.)

Duty Cycle The ratio of on to off time in repetitive operation, generally expressed as the on percentage of total cycle time.

Electron A negatively charged particle of an atom.

Electron Flow   Electric current. The movement of electrons under the influence of an electromotive force.

Emitter   A main current terminal of a transistor, also associated with the base terminal and its control current.

EMR   Electromechanical relay with coil and mechanical contacts.

Energize   Turn on; the application of control power.

Fast Recovery Diode   A diode which is specially processed to have a short recovery time; used as suppression diode across inductor to protect SSR with FET output.

FET   Field-effect transistor. Principle of operation differs from that of bipolar types. Voltage applied between gate and source terminals modulates the device resistance to current flow between drain and source terminals, by means of a field set up in the channel region.

Form, Contact   Contact or output arrangement, as in form A for single pole, single throw, normally open, etc.

Forward Direction   In semiconductors, the conducting, or low resistance direction, of current flow.

Frequency   Cycles per second (repetition rate).

Fuse   A protective device which melts and interrupts the current when its electrical rating is exceeded.

Gain   The ratio of an output to an input (voltage, current, or power) Beta: Greek letter used as a symbol for current gain.

Gate   Logic switching element (e.g., TTL). The control terminals of an FET or thyristor.

Heat Sink   A material with good heat conducting/dissipating properties to which an SSR is attached to cool and maintain its output junctions within the proper temperature range.

HEMR   Hybrid Electromechanical Relay. A combination of electronic and semiconductor devices in a relay with an electromechanical output (contacts).

Hertz (Hz)   Unit of frequency equal to one cycle per second.

Holding Current   The minimum (load) current required to maintain a thyristor in its conducting state.

HSSR   Hybrid solid-state relay. A combination of electronic and electromechanical devices in a relay with a solid-state output. A solid-state relay that uses a combination of differing semiconductor technologies (e.g., field effect/bipolar/IC).

Hybrid   A circuit combining different technologies.

Hysteresis   The difference in electrical levels at which a change in state will occur with increasing and decreasing electrical stimulus (e.g., EMR turn-on and turn-off levels).

Inductance   An electromagnetic property which can oppose a current change and also store a charge. The unit of inductance is the henry.

Input   The control side of an SSR to which a command signal is applied in order to effect a change in its output conditions.

Input Current (Maximum)   Current drain on the control source at specified SSR input voltages and on-off conditions.

Input Impedance/Resistance   Minimum effective SSR input resistance at a given voltage which defines input power and sensitivity.

Input Voltage   See turn-on/turn-off voltages.

Insulation Resistance (Minimum)   Resistive value usually measured at 500 volts DC, input to output, or both to case.

Insulator   A material which prohibits the easy flow of electric current.

Inverter   An electrical circuit that changes direct current to alternating current, or vice versa. An electrical circuit which has a positive (logic 1) output when the input is negative (logic 0), and vice versa.

Isolation   The value of insulation resistance, dielectric strength, and capacitance measured between the input and output, input to case, output to case, and output to output as applicable.

I^2t (Maximum)   Nonrepetitive pulse current capability of SSR given for fuse selection. Expressed as "ampere squared seconds" with typical half cycle pulse width.

Joule   A unit of work and energy in the MKS system (watt/seconds). Used as a measure of the transient energy capability of MOVs.

Junction   The region between semiconductor layers of opposite polarities.

Latching Current   The minimum initial load current required to cause a thyristor to remain in the conducting state immediately after switching and removal of trigger signal.

Latching Relay   A relay that maintains its last operated state after control power is removed.

Leakage Current   The maximum leakage current conducted through the SSR output terminals, in the off-state.

LED (Light-Emitting Diode)   Commonly used as the light-emitting source in a photocoupler.

Load Current (Maximum) The maximum steady-state load current capability of an SSR, which may be further restricted in use by the thermal dictates of heat sink and ambient temperature conditions.

Load Current (Minimum) The minimum load current required by the SSR to perform as specified.

Main Terminal Designated as "Main Terminal 1" and "Main Terminal 2" to identify the power terminals of a bidirectional thyristor (triac).

Make The closure of open contacts or conduction of SSR output to complete an electric circuit.

Make-Before-Break A "contact" closure that completes one electrical circuit before breaking another, thus causing a momentary short between circuits (form D).

Monolithic A semiconductor circuit having all components on the same piece of silicon.

Monostable A two-state device (relay) that will operate with applied control power and return to its off state when control power is removed.

MOSFET Metal oxide semiconductor field-effect transistor. The control electrode (gate) is electrically isolated from the source electrode by a layer of silicon oxide. A voltage applied between the gate and source will produce a current flow between drain and source.

MOV Metal oxide varistor, commonly used with AC SSRs to suppress bidirectional voltage transients. Has nonlinear voltage-dependent resistive characteristic that drops rapidly with increasing voltage.

Normally Open (N O) A contact or SSR output that is open when no control power is present at the input (coil). (Form A).

Normally Closed (N C) A contact or SSR output that is closed when no control power is present at the input (coil). (Form B).

Notch That period in an AC line cycle when a zero switching SSR is in a state that will allow turn-on (permissible switching window).

Off State Nonconducting (high) state of SSR output.

Off-State Voltage The maximum steady-state voltage that an SSR output can withstand without malfunction or damage. Usually specified as the upper limit of operating voltage.

Ohm Unit of measure of electrical resistance. The ohm is the resistance which allows the flow of one ampere when an electromotive force of one volt is applied.

Ohm's Law Relates resistance to current and to voltage. Resistance equals electromotive force (E) divided by current (I), $R = E/I$.

On State Conducting (low) state of the SSR output.

On-State Voltage (Maximum) The peak voltage that appears across the SSR output terminals at full rated load.

Operate A relay operates when it performs its prescribed switching function.

Operating Current A range of load currents including the minimum required to operate, and the maximum capability of the SSR. (See load current.)

Operating Voltage The range of voltages applied to the output, over which an SSR will continuously block or switch and otherwise perform as specified.

Output The load side of an SSR that performs the required switching function.

Overcurrent Maximum allowable SSR momentary current flow for a specific time duration. (Typically expressed as an RMS value for a one second duration.)

Peak Repetitive On Voltage The maximum off-state voltage that appears across the SSR output terminals immediately prior to turn-on at each half cycle, with control signal applied.

Phase Control Turn-on of a nonzero switching SSR (each half cycle), at a phase angle determined by the control signal source.

Photocoupler/Isolator A combination of light-emitting diode and light-sensitive semiconductor used to transmit information optically while providing electrical isolation. Commonly used isolating element for coupling the control signal to the output in an SSR.

Photovoltaic Photocoupling system whereby usable energy from the control signal traverses the gap for biasing purposes in the SSR output.

Pickup An EMR picks up when it changes from the unenergized condition to an energized condition. (See turn-on voltage.)

Pole, Double A contact arrangement that includes two separate contact combinations.

Pole, Single A contact arrangement in which all contacts connect in one position or another to a common contact.

Power Dissipation The maximum average power dissipation (watts) resulting from the effective voltage drop (power loss) in the SSR output semiconductor.

Pull-In See pickup (turn-on voltage).

Pulse A short time occurrence of voltage, current, or power.

Ratings The specified maximum performance capability of an SSR.

Reactance Opposition to the flow of AC current in both capacitive and inductive loads.

Recovery Time The time needed for a semiconductor to regain its normal blocking capability, after conduction (typically: diodes, SCRs etc.).

Rectifier A semiconductor used to convert AC power to DC. It allows current to flow in one direction (forward) and prevents the flow of current in the opposite direction (reverse).

Reed Relay Glass-enclosed contact assembly made of magnetic material and operated by external magnetic field.

Regulator A device (like a zener diode) whose function is to maintain a designated voltage or current.

Relay An electromechanical (EMR) or a solid-state (SSR) device used to switch electrical circuits by means of a comparatively low power, electrically isolated signal.

Release See dropout, turn-off voltage.

Resistance Defines the degree of limitation to the flow of electric current presented by circuit elements (measured in ohms).

RMS Voltage (Root-Mean-Square) The value of alternating voltage (AC) that would produce the same power dissipation as continuous voltage (DC) in a resistive load. For a sine wave, RMS is 0.707 times the peak value.

SCR Silicon controlled rectifier. Unidirectional semiconductor of the thyristor family with latching properties.

Semiconductor Device Transistors, diodes, etc., manufactured from semiconducting materials such as germanium and silicon.

Semiconductor Fuse A specially constructed fast-acting fuse capable of protecting semiconductor devices, with opening times typically less than 10 milliseconds.

Sensor A device used to sense changes in light, temperature, voltage, current, etc.

Snubber A resistor-capacitor combination placed across the SSR output terminals to control dv/dt and transients in thyristor circuits.

Solenoid A device that converts electrical energy into linear motion by means of a movable iron core within a coil (electromagnetic action).

Solid State A general term applied to a component, circuit, or system using semiconductors.

SSR (Solid-State Relay) Isolated on-off switch composed of nonmoving electrical parts (i.e., primarily semiconductors, transformers, and passive components).

Surge Current   The maximum allowable SSR momentary current flow for a specific time duration. (Typically specified as a peak value for one line cycle for AC.)

Temperature, Maximum Case ($T_{C \ (max)}$)   The maximum allowable case temperature for a given load current, measured at a specified point (°C).

Temperature, Maximum Junction ($T_{J \ (max)}$)   The maximum junction temperature of the output switching semiconductor (°C).

Thermal Resistance (R_θ)   Expressed in "degrees celsius per watt" (°C/W), this value defines the temperature gradient in the path between the power generated in the output SSR semiconductor and the final dissipating medium (heat sink/air).

Thyristor   A semiconductor bistable device comprising three or more junctions (PNPN, etc.). The generic name for a family of gate controlled switches including SCRs and triacs.

Time-Delay Relay   A relay in which operation or release is delayed. Accomplished internally by an electronic (RC) network.

Transient   Brief overvoltage overcurrent excursion from normal condition.

Transient Overvoltage   The maximum allowable brief excursion of applied voltage that an SSR can withstand without damage or malfunction while maintaining its off state.

Transistor   Generally a three-terminal semiconductor device with DC current flow between two terminals modulated by the third. A bipolar transistor is essentially a current-controlled device, while a field-effect transistor is a voltage-controlled device.

Triac   Bidirectional semiconductor of the thyristor family. Performance is similar to that of an inverse-parallel pair of SCRs, triggered by a single gate electrode.

Trigger   To turn on an SCR or triac at its gate.

TTL/T²L   Common bipolar IC logic family designated transistor-transistor logic, with "totem pole" (source/sink) output.

Turn-Off-Time (Maximum)   The time between removal of the control signal and the transition of the SSR to its fully off state (blocking).

Turn-Off-Voltage (Must Release)   The voltage applied to the SSR input at or below which the output must be in the off state (normally open).

Turn-On Time (Maximum)   The time between the application of a turn-on control signal and the transition of the SSR output to its fully conducting state.

Turn-On Voltage (Must Operate) The voltage applied to the SSR input at or above which the output must be in the on state (normally open).

Varistor See MOV

Volt Unit of electromotive force required to cause 1 ampere of current to flow through a 1-ohm resistor.

Watt The electrical unit of power; the product of volts and amperes.

Zener Diode (Also Avalanche Diode) Similar to a diode in the forward direction but it has an avalanche or breakdown characteristic at a specific voltage in the reverse direction. Used as a regulator, but more commonly in SSRs, as a unidirectional (DC) transient suppressor.

Zero Voltage Turn-On The maximum (peak) off-state voltage that appears across the SSR output terminals immediately prior to initial turn-on, following a turn-on control signal. (See notch.)

Appendixes

Appendixes

A

Additional Reference Material

Articles

Bachman, P. "Solid State Relay Applications." *Electronic Design*. October 25 1977.

Beddoe, S. "Interfacing Relays With Semiconductors." *New Electronics (UK)*. October 19 1977.

Beigel, J., McLeod, A.D. "Inductively Loaded SSRs Control Turn-On To Eliminate First Cycle Current Surges." *Electronic Design*. March 15 1979.

Bishop, A. "Applied Ideas—Phase Sequence Detector." *Electronics Engineering (UK)*. April 1 1977.

Bishop, A. "Beyond Usual Specs Is a Path to Better Relays." *Electronic Products*. August 1 1985.

Bishop, A. "Breaking the Noise Barrier." *E.E. Times*. August 1974.

Bishop, A. "Consideration and Application Tips for Designing With Solid State Relays." *E.E. Times-Magazine*. September 13 1976

Bishop, A. "Mini Glossary On Solid State Relays." *Electronic Products*. March 1979.

Bishop, A. "Modular I/O Interface Designs for Micros." *Design News*. November 3 1980.

Bishop, A. "Relays Control Electronic 'Muscle.' " *Instruments and Control Systems*. December 1977.

Bishop, A. "Relays That Fight EMI." *Machine Design*. January 20 1977.

Bishop, A. "Relays That Fight EMI," (Japanese). *Machine Design Japan*. May 1977.

Bishop, A. "Solid State I/O Converter Modules Feature Optically Isolated Inputs." *Canadian Electronics Engineering*. May 1976.

Bishop, A. "Solid State Relay Applications." *New Electronics (UK)*. May 17 1977.

Bishop, A. "Usage of Semiconductor Relays" (German). *Electronik Industrie*, Germany. March 28 1983.

Bishop, A. "What To Look For In SSRs." *Instruments and Control Systems*. April 1977.

Borrell, S. "Designing With Solid State Relays." *Electron (UK)*. July 21 1977.

Collins, W., Kinzer, D.M. "Solid-State Relay Outperforms Reeds for Small Analog Signals." *Electronic Design*. March 8 1984.

Collins, H.W. "Solid State Relays, Where Are We Going?" *New Electronics (UK)*. June 22 1975.

"Controlling Transformer Inrush Currents." *EDN*. July 1966.

Cooper, P. "The Solid State of The Solid-State Relay Industry." November 21 1983.

Coughlin, V., Sclater, N. "Solid State Relays Meet OEM Requirements." *Design Engineering*. March 1981.

Craney, Patrick M. "Electromechanical Vs. Solid-State Relays." *Electronic Products*. March 28 1983.

Emery, J. "MOSFET-Based Switches Compete with EMRs." *Electronic Design*. June 13 1985.

Gibbs, S. "Effects of Loads and Transients on Solid State Relays." *NARM Proceedings*. April 1972.

Gross, T.A.O. "Solid State Control for Capacitor-Start Motors." *Machine Design*.

Grossman, M. "Focus On Relays." *Electronic Design*. December 20 1978.

Hildrew, S. "Triac Testing for SSR Applications." *New Electronics (UK)*. December 12 1978.

Johnson, P. "Solid-State Relays—A Guide to Design." *EDN* Oct. 5 1973.

Lidow, A., Collins, W. "Solid State Power Relays Enter the IC Era." *Electronics*. December 29 1982.

Lord, P. "Components Update: Solid State Relays." *EE Times*. August 15 1983.

Marketing/Engineering Staff. "Solid State Relays." *Teledyne Relays EE Times*. June 8 1981.

Marrin, K. "I/O Boards and Software Give μCs Remote Control over Solid-State Relays." *EDN*. May 30 1985.

McGowan, M.J. "Remote I/O Systems Simplify Software While Multiplexing Signals." *Control Engineering*. May 1 1980.

Mead, T. "New Vistas May Be Opening For The Solid State Relay." *Electronic Business*. February 1982.

Moore, D.W. "Solid State Relays." *Electron (UK)*. July 17 1978.

Moore, D.W. "The Solid State Relay As A Power Switch." *Electron (UK)*. May 22 1975.

Moore, D.W. "Using The Solid State Relay." *Electron (UK)*. December 12 1977.

Morris. R.M. "Current Surges In Circuits Using Zero-Voltage Switching." *New Electronics (UK)*. March 8 1977.

Motto, J.W. "A New Quantity to Describe Power Semiconductor Sub-cycle Current Ratings." *IEEE Transactions*, Vol IGA-7, #4 July/August 1971.

"New Generations Set For I/O Market." *Electronic News*. April 20 1981.

Pancake, J.R. "Solid State Relays Enhance Computer Control." *Instruments and Control Systems*. June 1980.

"Quick Guide to Solid-State Relays." *Beep*. February 1985.

Rand, M.B. "Despite Sleepy Image, Relay Technology Makes Strides Forward." *Electronics Week*. June 10 1985.

Sahm, W. III. "Solid State Relays Aren't All Alike." *Electronic Products*. July, 15, 1974.

Schneider, S. "The Growing Family of Solid-State Relays." *Machine Design*. 1979.

Schreier, P.G. "Optocouplers and Power Semiconductors Initiate Quiet Revolution In SSRs." *EDN*. September 30 1981.

"Solid State Relays Are Fast, Controllable." *Product Engineering*. March 1979.

"SSRs Vs. Reeds Vs. EMRs." *Evaluation Engineering*. May/June 1981.

"Switching Module,s" Reference Issue. *Machine Design*. May 17 1979.

Teschler, L. "Modular I/O For Microcomputers." *Machine Design*. November 9 1978.

Teschler, L. "Remote I/O For Industrial Control." *Machine Design*. July 9 1981.

Thomas, E.U. "The Great Zero Cross-Over Hoax." *NARM Proceedings*. May 1974.

Voiculescu, A. "Progress in Solid State Relays." *Electronic Products*. June 1980.

Weghorn, F. "Input/Output Modules Refine Industrial Control." *Electronic Products*. 1981.

Weghorn, F. "Solid State Relays." *Electronic Products*. October 1981.

Application Notes

"Beware of Zero-Crossover, Switching of Transformers." P.M. Craney. Potter & Brumfield.

"Fuse-Thyristor Coordination Primer." B. Botos and B. Haver. AN-568 Motorola Semiconductor Products.

"Handbook of Solid State Relays." Grayhill.

"Introduction to Solid State Relays." R.W. Fox. IR Crydom.

"Pitfalls To Avoid In Selecting I/O Modules." P.M. Craney. Potter & Brumfield.

"Selecting Solid State Relays for Appliance Applications." N. Yudewitz. Magnecraft.

"Semiconductor Fuses." Bulletin SCF. Bussmann Div. McGraw Edison Co. 1981.

"Solid State Relay Applications Handbook." A. Bishop. Teledyne Relays 1976.

"Solid State Relay Application Tips." Opto-22.

"Solid State Relays." Magnecraft.

"Solid State Relays for AC Power Control." T. Malarkey. Motorola 1975.

"Solid State Users Handbook." Potter & Brumfield.

Books

"Alternating Current Machines." Halstead Press, John Wiley and Son.

"Electric Fuses." A. Wright/P.G. Newbery. IEE (UK) Power Engineering, Series 2.

"Engineer's Relay Handbook." NARM, 1966.

"Heat Sink Application Handbook." J. Spoor. AHAM 1974.

"Power Semiconductor Databook." SDB-2 International Rectifier 1985.

"SCR Applications Handbook," 2nd Edition. International Rectifier 1977.

"SCR Designers Handbook." Westinghouse Electric Corp. 1970.

"SCR Manual (6th Edition)." General Electric Company 1979.

"Semiconductor Fuse Applications Handbook." International Rectifier 1972.

"Thyristors-Rectifiers." Electronics Data Library. General Electric Company 1982.

"Transient Voltage Suppression", 4th Edition. General Electric Company.

Test Specifications

NEMA ICS2-230	Electrical Noise Immunity Test
ANSI C37.90-1971 ANSI C37.90a-1974 IEEE STD 313-1971 IEEE STD 472-1974	Guide for Surge Withstand Capability (SWC) Tests
EIA/NARM RS-443-1979	Standard for Solid-State Relays
UL 508	Safety Standard for Industrial Control Equipment
VDE 0806 IEC 380	Safety of Electrically Energized Office Machines
VDE 0660 IEC 158-2	Low Voltage Switchgear and Control Gear (Solid-State Contactors)
VDE 0871 VDE 0872 VDE 0875	Radio Interference Regulations (Europe)
FCC Docket 220780	Radio Interference (EMI) Technical Standard for Computing Equipment (USA)
IEC 435	Safety of Data Processing Equipment

B

Useful Data

Commonly Used Formulas

Inductive Loads

Inductive reactance (ohms): $X_L = 2\pi f L$

Capacitive reactance (ohms): $X_C = \dfrac{1}{2\pi f C}$

Power factor: $\cos \theta = \dfrac{R}{Z} = \dfrac{W}{VA}$

Impedance (ohms): $Z = \sqrt{R^2 + X_L^2} = \dfrac{R}{\cos \theta} = \dfrac{X_L}{\sin \theta}$

$$R = Z \cos \theta, \quad X_L = Z \sin \theta$$

Voltage to phase angle (deg): $\theta = \sin^{-1} \dfrac{E(pk)}{\text{Line } E\,(\sqrt{2})}$

Phase to angle to time (ms): $tx = \theta \dfrac{0.5 \text{ cycle (ms)}}{0.5 \text{ cycle (deg)}}$

where
    π = 3.14
    f = hertz
    L = henrys
    C = farads
    θ = degrees
    Z = ohms
    R = ohms

Snubbers

$$C = \dfrac{4L}{R^2}, \quad R = 2\propto\sqrt{L/C}, \quad \dfrac{dv}{dt} = \dfrac{ER}{L}$$

where
    C = μF
    L = μH
    R = ohms
    $\dfrac{dv}{dt}$ = V/μs
    $\propto$ = damping factor (0.7 typ)

Fusing

Amp-squared-seconds: $I^2t = \left(\dfrac{I^2_{pk(surge)}}{2} \right).0083 \text{ seconds}$

Subcycle surge from I^2t and t_x: $I_{pk(surge)}\sqrt{\dfrac{I^2t}{t_x}} \times 2$

t_x from I^2t and surge current: $t_x = \dfrac{I^2t}{I^2_{RMS}}$ or $2 \times \dfrac{I^2t}{I^2_{pk}}$

Clearing time (t_c) from I^2t and peak let-through current (I_{plt})

$$t_c = \dfrac{3I^2_t}{I^2_{plt}}$$

where
    I = amps
    t = time in seconds

Heat Sinking

Thermal resistance (°C/watt): $R_\theta = \dfrac{\Delta T}{P_{diss}}$

ΔT junction to ambient: $T_J - T_A = P_{diss}(R_{\theta JC} + R_{\theta SA})$

R_θ junction to ambient: $R_{\theta JA} = \dfrac{T_J - T_A}{P_{diss}}$

where
T = temperature (°C)
ΔT = temperature difference between two points
P_{diss} = watts
R_θ = thermal resistance (°C/W)
$R_{\theta JC}$ = junction-case
$R_{\theta CS}$ = case-sink
$R_{\theta SA}$ = sink-amb.

Switch and Logic Symbols

Shown here are some switch symbols commonly used for EMR and electronic functions that are sometimes used symbolically to describe SSR outputs and configurations.

DESCRIPTION	ANSI SYMBOL	IEC AND JIC SYMBOLS	LOGIC OR OTHER SYMBOLS
FORM A SPST-N O			
FORM B SPST-N C			
FORM C SPDT			

DESCRIPTION	ANSI SYMBOL	IEC AND JIC SYMBOLS	LOGIC OR OTHER SYMBOLS
2 FORM A (Series) DPST-N O			"AND"
2 FORM A (Parallel) DPST-N O			"OR"
2 FORM B (Parallel) DPST-N C			"NAND"
2 FORM B (Series) DPST-N C			"NOR"

Greek Alphabet

NAME	CAPITAL	LOWER CASE	DESIGNATES
Alpha	A	α	Angles. Area. Coefficents
Beta	B	β	Angles. Flux density. Coefficients
Gamma	Γ	γ	Conductivity. Specific gravity
Delta	Δ	δ	Variation. Density
Epsilon	E	ϵ	Base of natural logarithms
Zeta	Z	ζ	Impedance. Coefficients. Coordinates
Eta	H	η	Hysteresis coefficient. Efficiency
Theta	Θ	θ	Temperature. Phase angle
Iota	I	ι	Unit vector.
Kappa	K	κ	Dielectric constant. Susceptibility
Lambda	Λ	λ	Wave length
Mu	M	μ	Micro. Amplification factor. Permeability
Nu	N	ν	Reluctivity
Xi	Ξ	ξ	Co-ordinates

NAME	CAPITAL	LOWER CASE	DESIGNATES
Omicron	O	o	. . . .
Pi	Π	π	3.1416 (Ratio of circumference to diameter)
Rho	P	ρ	Resistivity
Sigma	Σ	ς	Sign of summation
Tau	T	τ	Time constant. Time phase displacement
Upsilon	Υ	υ	. . . .
Phi	Φ	φ	Magnetic flux. Angles
Chi	X	χ	Electric susceptibility. Angles
Psi	Ψ	ψ	Dielectric flux. Phase difference
Omega	Ω	ω	Capital, ohms. Lower case, angular velocity

Mathematical Constants

$$\pi = 3.14$$
$$2\pi = 6.28$$
$$(2\pi)^2 = 39.5$$
$$4\pi = 12.6$$
$$\pi^2 = 9.87$$
$$\frac{\pi}{2} = 1.57$$
$$\frac{1}{\pi} = 0.318$$
$$\frac{1}{2\pi} = 0.159$$
$$\frac{1}{\pi^2} = 0.101$$
$$\frac{1}{\sqrt{\pi}} = 0.564$$

$$\sqrt{\pi} = 1.77$$
$$\sqrt{\frac{\pi}{2}} = 1.25$$
$$\sqrt{2} = 1.41$$
$$\sqrt{3} = 1.73$$
$$\frac{1}{\sqrt{2}} = 0.707$$
$$\frac{1}{\sqrt{3}} = 0.577$$
$$\log \pi = 0.497$$
$$\log \frac{\pi}{2} = 0.196$$
$$\log \pi^2 = 0.994$$
$$\log \sqrt{\pi} = 0.248$$

Mathematical Symbols

$\times$ or $\cdot$ Multiplied by
$\div$ or :   Divided by
$+$ Positive. Plus. Add
$-$ Negative. Minus. Subtract

Mathematical Symbols

$\pm$		Positive or negative. Plus or minus
$\mp$		Negative or positive. Minus or plus
$=$	or ::	Equals
$\equiv$		Identity
$\cong$		Is approximately equal to
$\neq$		Does not equal
$>$		Is greater than
$\gg$		Is much greater than
$<$		Is less than
$\ll$		Is much less than
$\geq$		Greater than or equal to
$\leq$		Less than or equal to
$\therefore$		Therefore
$\angle$		Angle
Δ		Increment or Decrement
$\perp$		Perpendicular to
$\parallel$		Parallel to
$\lvert n \rvert$		Absolute value of n

Conversion Table

TO CONVERT	INTO	MULTIPLY BY
ampere hours	coulombs	3,600.0
ampere turns	gilberts	1.257
atmosphere	cms of mercury	76.0
atmosphere	in. of mercury (at 0°C)	29.92
Btu	ergs	1.0550×10^{10}
Btu	ft-lbs	778.3
Btu	kilogram-calories	252.0
Btu/sq ft/min	watts/sq in.	0.1221
Centigrade	Fahrenheit	$(C° \times 9/5) + 32$
centimeter-grams	pound-feet	0.01
circular mils	sq cms	5.067×10^{-6}
circular mils	sq inches	7.854×10^{-7}
coulombs	Faradays	1.036×10^{-5}
cubic feet/min	gallons/sec	0.1247
degrees (angle)	radians	0.01745
dynes	grams	1.020×10^{-3}
dynes	joules/cm	10^{-7}
dynes	poundals	7.233×10^{-5}
ergs	Btu	9.480×10^{-11}
ergs	dyne-centimeters	1.0
ergs	foot pounds	7.367×10^{-8}
ergs	joules	10^{-7}
faradays	ampere hours	26.80

TO CONVERT	INTO	MULTIPLY BY
faradays	coulombs	9.649×10^4
feet of water	atmospheres	0.02950
foot pounds	hp-hrs	5.050×10^{-7}
foot pounds	kilowatt-hrs	3.766×10^{-7}
gallons	cubic feet	0.1337
gallons	cubic meters	3.785×10^{-3}
gallons (liq. Br Imp)	gallons (U.S. liq)	1.70095
gausses	lines/sq in	6.452
gausses	webers/sq cm	10^{-8}
gram-calories	Btu	3.9683×10^{-3}
joules	Kg-calories	2.388×10^{-4}
joules	Kg-meters	0.1020
joules	watt-hours	2.778×10^{-4}
joules	foot-pounds	0.7376
joules/cm	grams	1.120×10^4
joules/cm	poundals	723.3
joules/cm	pounds	22.48
kilowatt hours	Btu	3,413
kilowatt hours	ergs	3.600×10^{13}
kilowatt hours	horsepower-hours	1.341
knots	feet/hr	6,080
knots	nautical miles/hr	1.0
league	miles (approx.)	3.0
liters	gallons (U.S. liq.)	0.2642
maxwells	kilolines	0.001
maxwells	webers	10^{-8}
ounces (fluid)	cu inches	1.805
ounces (troy)	ounces (avdp.)	1.09714
quadrants (angle)	degrees	90.0
quadrants (angle)	minutes	5,400.0
quadrants (angle)	radians	1.571
quadrants (angle)	seconds	3.24×10^5
radians	degrees	57.30
radians/sec	revolutions/min	9.549
square centimeters	circular mils	1.973×10^5
square feet	acres	1.076×10^{-3}
square inches	circular mils	1.273×10^6
temperature (°C) + 273	absolute temperature (°C)	1.0
temperature (°C) + 17.78	temperature (°F)	1.8
temperature (°F) + 460	absolute temperature (°F)	1.0
tons (long)	kilograms	1.016
tons (short)	kilograms	1.120
watts	Btu/hr	3.413
watts	ergs/sec	10^7
watts	horsepower	1.341×10^{-3}
yards	meters	0.9144

Temperature Conversion

Centigrade

$$°C = \frac{5}{9}(°F - 32)$$

Fahrenheit

$$°F = \left(\frac{9}{5} \times °C\right) + 32°$$

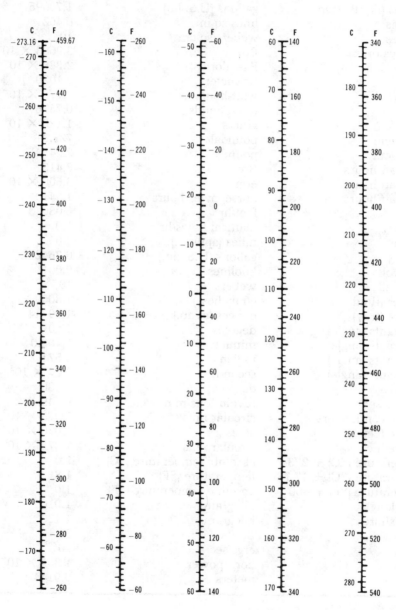

Thermal Properties of Commonly Used Materials

MATERIAL	CONDUCTIVITY WATTS/IN °C	RESISTIVITY °C IN/WATT
Air (100°C)	0.001	1,000
Aluminum	5.63	0.178
Alumina (Al Oxide)	0.55	1.82
Brass	2.97	0.337
Beryllium	4.02	0.249
Beryllia (Be Oxide)	5.15	0.194
Copper	9.93	0.101
Duralumin	4.07	0.246
Epoxy (Conductive)	0.02	50.0
German Silver	0.81	1.23
Gold	7.54	0.133
Iron (Pure)	1.90	0.526
Lead	0.88	1.14
Molybdenum	3.70	0.270
Nichrome	0.33	3.03
Nickel	1.52	0.658
Nickel Silver	0.84	1.19
Phosphor Bronze	1.80	0.555
Platinum	1.75	0.571
Silicon	2.13	0.469
Silver	10.6	0.094
Steel (1045)	1.27	0.787
Steel, Stainless (347)	0.41	2.44
Tin	1.60	0.625
Zinc	2.87	0.348

$$\text{Thermal Resistance } (\theta) \text{ °C/W} = \frac{\text{Resistivity (°C in/w)} \times \text{Thickness (in)}}{\text{Area (sq in)}}$$

Metric Prefixes

FACTOR	PREFIX	SYMBOL	FACTOR	PREFIX	SYMBOL
10^{18}	exa	E	10^{-1}	deci	d
10^{15}	peta	P	10^{-2}	centi	c
10^{12}	tera	T	10^{-3}	milli	m
10^{9}	giga	G	10^{-6}	micro	μ
10^{6}	mega	M	10^{-9}	nano	n
10^{3}	kilo	k	10^{-12}	pico	p
10^{2}	hecto	h	10^{-15}	femto	f
10^{1}	deka	da	10^{-19}	atto	a

Table of Standard Annealed Bare Copper Wire

AWG B & S GAUGE	DIAMETER IN MILS	CROSS SECTION		LBS PER 1000 FT	FT PER LB	OHMS PER 1000 FT AT 20° C	FT PER OHM AT 20° C	OHMS PER LB AT 20° C	MAX CURRENT (AMPS) INSULATED
		Circular Mils	Square Inches						
0000	460.0	211,600	0.1662	640.5	1.561	0.04901	20,400	0.00007652	225
000	409.6	167,800	0.1318	507.9	1.968	0.06180	16,180	0.0001217	175
00	364.8	133,100	0.1045	402.8	2.482	0.07793	12,830	0.0001935	150
0	324.9	105,500	0.08289	319.5	3.130	0.09827	10,180	0.0003076	125
1	289.3	83,690	0.06573	253.3	3.947	0.1239	8,070	0.0004891	100
2	257.6	66,370	0.05213	200.9	4.977	0.1563	6,400	0.0007778	90
3	229.4	52,640	0.04134	159.3	6.276	0.1970	5,075	0.001237	80
4	204.3	41,740	0.03278	126.4	7.914	0.2485	4,025	0.001966	70
5	181.9	33,100	0.02600	100.2	9.980	0.3133	3,192	0.003127	55
6	162.0	26,250	0.02062	79.46	12.58	0.3951	2,531	0.004972	50
7	144.3	20,820	0.01635	63.02	15.87	0.4982	2,007	0.007905	
8	128.5	16,510	0.01297	49.98	20.01	0.6282	1,592	0.01257	35
9	114.4	13,090	0.01028	39.63	25.23	0.7921	1,262	0.01999	
10	101.9	10,380	0.008155	31.43	31.82	0.9989	1,001	0.03178	25
11	90.74	8,234	0.006467	24.92	40.12	1.260	794	0.05053	
12	80.81	6,530	0.005129	19.77	50.59	1.588	629.6	0.08035	20
13	71.96	5,178	0.004067	15.68	63.80	2.003	499.3	0.1278	
14	64.08	4,107	0.003225	12.43	80.44	2.525	396.0	0.2032	15
15	57.07	3,257	0.002558	9.858	101.4	3.184	314.0	0.3230	
16	50.82	2,583	0.002028	7.818	127.9	4.016	249.0	0.5136	6
17	45.26	2,048	0.001609	6.200	161.3	5.064	197.5	0.8167	
18	40.30	1,624	0.001276	4.917	203.4	6.385	156.6	1.299	3
19	35.89	1,288	0.001012	3.899	256.5	8.051	124.2	2.065	
20	31.96	1,022	0.0008023	3.092	323.4	10.15	98.50	3.283	
21	28.46	810.1	0.0006363	2.452	407.8	12.80	78.11	5.221	
22	25.35	642.4	0.0005046	1.945	514.2	16.14	61.95	8.301	
23	22.57	509.5	0.0004002	1.542	648.4	20.36	49.13	13.20	
24	20.10	404.0	0.0003173	1.223	817.7	25.67	38.96	20.99	
25	17.90	320.4	0.0002517	0.9699	1,031.0	32.37	30.90	33.37	
26	15.94	254.1	0.0001996	0.7692	1,300	40.81	24.50	53.06	
27	14.20	201.5	0.0001583	0.6100	1,639	51.47	19.43	84.37	
28	12.64	159.8	0.0001255	0.4837	2,067	64.90	15.41	134.2	
29	11.26	126.7	0.00009953	0.3836	2,607	81.83	12.22	213.3	
30	10.03	100.5	0.00007894	0.3042	3,287	103.2	9.691	339.2	
31	8.928	79.70	0.00006260	0.2413	4,145	130.1	7.685	539.3	
32	7.950	63.21	0.00004964	0.1913	5,227	164.1	6.095	857.6	
33	7.080	50.13	0.00003937	0.1517	6,591	206.9	4.833	1,364	
34	6.305	39.75	0.00003122	0.1203	8,310	260.9	3.833	2,168	
35	5.615	31.52	0.00002476	0.09542	10,480	329.0	3.040	3,448	
36	5.000	25.00	0.00001964	0.07568	13,210	414.8	2.411	5,482	
37	4.453	19.83	0.00001557	0.06001	16,660	523.1	1.912	8,717	
38	3.965	15.72	0.00001235	0.04759	21,010	659.6	1.516	13,860	
39	3.531	12.47	0.000009793	0.03774	26,500	831.8	1.202	22,040	
40	3.145	9.888	0.000007766	0.02993	33,410	1049.0	0.9534	35,040	

Index

A

AC SSR(s). *See also* Solid-state relays
    (SSRs)
    circuits, 28-30, 33-35, 37-38, 73
    motor ratings, 88
    response time for, 95
    switches, 20-23
    thyristors used with, 15, 34, 95
Analog converter module, 128
Antex Electronics Corp., 179, 187
Applied Electrotechnology, Int., 187

B

Bidirectional switching of circuits,
    134
Bipolar devices, 141-42, 149, 155
Bounce suppression, special function
    switching with, 144-45
Buffered output modules for
    microcomputers, 126-27

C

Canadian Standards Association
    (CSA), 91
Capital Power Devices Ltd., 187
Celduc, 187
Circuit configurations, 43, 58, 82-84

CMOS logic system, 4, 55, 126-27
Compatible chassis mountable SSRs
    with dual SCR outputs, table of,
    177
    with triac output, table of, 178
Complementary power switching,
    132
Continental Industries, Inc., 187
Coupling system, 30. *See also* Optical
    coupling
Crydom controls, 1
Crydom Division International
    Rectifier, 177-78, 180-82, 187
Current
    ambient temperature versus, 64
    input range, 43-44, 161-62
    instant flow of, 85, 93
    off-state leakage of, 163-64
    output range, 45-46
    SSR failure caused by surge
    of, 72

D

Darlington output switching device,
    18-19, 52
Data processing equipment, safety
    of, 206
DC SSR(s). *See also* Solid-state relays
    (SSRs)
    AC converted to, 51-52

DC SSR(s)—*cont.*
    circuits, 27-28, 31-33, 35, 74
    motor ratings, 87-88
    reversing motor drive for, 154
    switches, 16-20
Dielectric strength testing for SSRs, 46, 160-61
Diodes as an effective suppressor, 103-6, 152
Dionics, Inc., 187
Douglas Randall Div. of Walter Kidde & Co., 178, 187
Drive source for ICs, 55-56

E

Electrical noise immunity test, 206
Electrol, Inc., 178, 180, 182, 187
Electromagnetic interference (EMI), 5, 32, 81, 85-86, 94
Electromatic, 187
Electromechanical relay (EMR), 1
    advantages of the, 4
    combined with SSR, 135
    compared to SSR, 5-7, 131
    disadvantages of the, 4
    hybrid, 2
    hysteresis in an, 30
    line short unlikely with an, 149
Electronic Instrument & Specialty Corporation, 188

F

Field effect (FET) operating-current threshold, 27-28
FR Electronics, 188
Fuses for semiconductors, 112-20
Fusing, 149

G

General parameters as characteristics of SSR, 46-47

Gordos International Corp., 177-78, 180-81, 188
Gould, Inc. Industrial Controls Division, 188
Grayhill, Inc., 179, 188
Guardian Electric Manufacturing Company, 179, 181, 188
Guide for surge withstand capability (SWC) tests, 206

H

"Hall effect" coupling device, 9
Halogen cycle lamps, 91
Hamlin, Inc., 178, 181, 188
Hasco Components, Inc., 188
Hazardous situations for user or equipment, 148-50
Heat sinking for temperature measurement of SSR, 66-70
Heinemann Electric Co., 188
Hybrid electromechanical relay (HEMR), 2
Hybrid solid-state relay (HSSR), 1-3, 135
Hysteresis
    DC SSR incorporating of, 31-32
    requirement of, 30-31
    transformer curve, 77-78

I

Idec Syst. & Contr. Corp., 188
Inductive load switching, 75-76
Industrial control equipment, safety of, 206
Input/output (I/O) module packages, 1-3, 121-30
Input parameters as characteristics of SSR, 43-45
Integrated circuits (IC) drive source, 55-56
ITT Components, 189

K

Kipco Corp., 189

L

Latching SSR circuits, 132-33
Leach Corporation Relay Division,
     189
Leakage from the drive source, 56-57
Light-emitting diode (LED)
     DC input circuits used with, 27
     installation of the, 124
     optocoupling methods, 10-11, 30
"Light-pipe" method of coupling, 12
Line frequency for circuits, 29
Lineal feet per minute (LFM), 68
Low voltage switchgear and control
     gear (solid-state contractors),
     206

M

Magnecraft Electric Company, 177,
     189
Master Electronic Controls, 189
Mean time between failure (MTBF),
     173
Medium power SSRs with fast-on
     terminals and triac output,
     table of, 179
Metal oxide field-effect transistors
     (MOSFET), 15
     paralleling SSRs with, 149, 155
     for switching low-level signals,
     36
Metal oxide varistors (MOVs), 89,
     102, 106-12
Microcomputers, I/O interface
     modules for, 121-30
Microelectronic Relays, Inc., 189
Midland Ross Corp. Midtex Division,
     179, 189
Miniature (DIP) SSRs with dual SCR
     output, table of, 182

Motor starter switch, 135-36
MSI Electronics, Inc., 189
Multivibrator functions, 147-48

N

NMOS logic system, 4, 55, 126
Noise
     susceptibility, responsibility of
          manufacturer for, 93-95
     parasitic phototransistor, 94-95
NPN transistors, 16-20, 53-54

O

Office machines, safety of, 206
Omega Engineering, Inc., 189
Omron Electronics, Inc., 189
Optical (photo) coupling.
          See Optocoupling
Optocoupling
     input to output isolation
          capabilities, 9-10, 30
     photoresistor method of, 11-12,
          38
     phototransistor method of, 10-12,
          31, 33
Opto-22, 1, 177, 180-81, 189
Output parameters as characteristics
     of SSR, 45-46
Over/under voltage sensor, 138-39

P

Peak repetitive on voltage (PROV),
     150
Peripheral interface adapter (PIA),
     127
Phase-controlled dimming, 146-47
Phase-sequence detector, 140-41
Philips ECG, Inc., 189
Photoresistor as a optocoupling
     method, 11-12, 38

Photo-SCR
    correcting parasitic noise in,
        94-95
    as a coupling method, 11-12,
        35-38
Phototransistor
    correcting parasitic noise in,
        94-95
    as a coupling method, 10-12,
        31, 33
Plug-in I/O modules (triac and
        transistor outputs), table of,
        180
PNP transistors, 16-20, 53
Potter & Brumfield Division of AMF,
        Inc., 178-80, 190

R

Radio interference (EMI) technical
        standard for computing
        equipment (USA), 206
Radio interference regulations
        (Europe), 206
Random turn-on switch, 80-83, 166

S

Safety of equipment, 206
Semiconductor(s), 1-2
    fuses for, 112-20
    leakage current in the driving,
        53-56
    parameters, 42
    piezoelectric device, 9
    switches, 20
    thermal relationship between
        surrounding ambient and,
        60-61
Siemens Corporation, 190
Sigma Instruments, Inc., 179, 182,
        190
Silicon controlled rectifiers (SCR)
    bridge circuit, 25

(SCR)—*cont.*
    as commonly used output
        devices, 20-23
Silicon Power Cube, 190
Single-pole double-throw (SPDT)
        device, 58, 144
Single-pole single-throw normally
        open (SPST-N O) output
        configurations, 43, 58
Single-pole single-throw (SPST)
        device, 6, 58, 123, 131
Small plug-in SSRs with triac and
        transistor outputs, table of,
        181
Snubber network
    a major factor in transient
        voltage and dv/dt
        suppression, 98-102
    used with output switching
        devices, 25, 34
Solid-state relay(s) (SSR)
    advantages of, 3-5
    applications, 131-58
    characteristics, 41-50
    compared to EMR, 5-7
    coupling methods for, 9-14
    definition of, 1
    directory of manufacturers and
        suppliers of, 176, 187-90
    disadvantages of, 3-5
    drive methods, 51-58
    hybrid versions of, 1-3, 135
    introduction to, 1-9
    lamp switching for, 89-91
    motor switching for, 84-89
    operation, 27-40
    output switching devices for,
        15-26
    package styles of, 176, 186
    protective measures for, 93-120
    selecting the proper, 47-50
    standard for, 206
    surge ratings for, 71-92
    testing performance of, 159-74
    thermal design as a major factor
        in the use of, 59-70
    time-delay, 3

(SSR)—*cont.*
    users of, 3
Split phase motors, reversing motor
    drive for, 136-37
Standard for solid-state relays, 206
Static dv/dt measurement, 95-96,
    171-72
Struthers-Dunn, Inc., 190
Suppressors used where overvoltage
    transients occur, 102-3
Surge ratings, 72-75, 115
Surge-reducing capabilities, 81-82, 85
Surge withstand capability (SWC),
    110, 206
Switching dual supplies, 152-53
Switching techniques for highly
    inductive loads, 80-84

T

Teccor, 190
Teledyne Solid State Products, 182,
    190
Temperature control systems, 145-46
Terminal style for relays, 41-43
Thermal design, 59-70
    manufacturers' ratings for, 64
    relationships between
        semiconductor and
        surrounding ambient, 60-63
    resistance versus surface area,
        59-63, 65, 68-69
Theta-J, 182, 190
Three-phase motor reversal, 150-51
Three-phase switch for three-wire
    system, 140
Thyristor(s)
    used with AC switching, 15,
        34, 95
    driving high powered, 141-42
    a family of semiconductor
        switches, 20
    output configuration, 23, 149,
        155
    rate effect in, 95
    technology, 25

Thyristor(s)—*cont.*
    unidirectional, 22
    voltage threshold of, 150
Time-delay electromechanical relay
    (TDEMR) capabilities, 3
Time delay functions, 147-48
Time-delay solid-state relay
    (TDSSR), 3
Toshiba America, Inc. Electronic
    Components Division, 190
Transformer coupling
    as a direct control signal drive
        feature of, 13-14
    input to output capabilities of,
        9-10, 37
Transformer switching, 76-80, 156-57
Transient overvoltage, 165
Transient suppressors effectiveness,
    95
Transistor(s)
    used with AC/DC switching, 15
    DPDT switch from single, 142-44
    metal oxide field-effect
        (MOSFET), 15
    NPN, 16-20, 53-54
    outputs, 180-81
    PNP, 16-20, 53
    voltage, 18
Triac device, 22, 178-81
    output, 37-38, 96-97
    versus SCR, 24-25
TTL gate drive method of SSRs,
    53-55
Tungsten halogen lamps, 91

V

Voltage
    input range of, 43-44
    isolation, 46
    dependent nonlinear resistor,
        107
    operating load current ranges
        and, 170-71
    output range of, 45-46, 51-52
    over/under sensor, 138-39

SCR, 21
selecting the MOV for, 109
snubbers a major factor in
    transient, 98-102
switch gear and control gear, 206
transistor, 18
turn-on, turn-off, 161-63, 168-69
waveforms, 77, 82-84

Z

Zener diode, a low voltage
    suppressor, 103-6, 152
Zero crossing as used in AC SSRs,
    32-34, 36-38
Zero current turn-off feature, 77
Zero voltage turn-on feature, 76,
    80-83, 90, 137, 155, 166-68

MORE
FROM
SAMS

☐ Audio IC Op-Amp Applications (3rd Edition)   *Walter G. Jung*

This updated version of a classic reference will be welcomed by recording and design engineers and hobbyists using audio signal processing. This new edition covers the changes that have marked the Op-Amp field over the last few years and includes new devices such as the OP-27/37 and application ICs for automobile stereo and audio testing. The update also includes new applications circuitry to illustrate current usage, among them differential input/output IC devices. Jung is a recognized expert in his field and is the author of the definitive *IC Op-Amp Cookbook*.
ISBN: 0-672-22452-6, $17.95

☐ CMOS Cookbook (2nd Edition)

Don Lancaster

Don Lancaster is back—and IC design engineers and electronics hobbyists will be delighted! This revision of one of his best-selling titles retains its cookbook recipe for a handy but comprehensive reference for CMOS, the most popular and widely used digital logic family. Presented in the author's engaging writing style, this new edition includes the latest in integrated circuits, such as the 74HC series, as well as the older 4000 series and user-programmable CMOS devices such as EPROMs, PALs, and PLAs.
ISBN: 0-672-22459-3, $16.95

☐ Handbook of Electronics Tables and Formulas (6th Edition)

Howard W. Sams Engineering Staff

The latest edition of this useful handbook contains all of the formulas and laws, constants and standards, symbols and codes, service and installation data, design data, and mathematical tables and formulas you would expect to find in this reference standard for the industry. New formulas include power units, graphical reactance relations, power triangle, and decibels/voltage/power diagram. Also featured are computer programs (written for Commodore 64®, with conversion information for Apple®, Radio Shack, and IBM®) for calculating many equations and formulas.
ISBN: 0-672-22469-0, $19.95

☐ The Home Satellite TV Installation and Troubleshooting Manual

Frank Baylin and Brent Gale

For the hobbyist or electronics buff, this title provides a comprehensive introduction to satellite communication theory, component operation, and the installation and troubleshooting of satellite systems, including the whys and wherefores of selecting satellite equipment. The authors are respected authorities and consultants to the satellite communication industry. If you are among the 100,000 people per month who are installing a satellite system, you'll want to have this book in your reference library.
ISBN: 0-672-22496-8, $29.95

☐ IC Op-Amp Cookbook (3rd Edition)

Walter G. Jung

Hobbyists and design engineers will be especially pleased at this new edition of the industry reference standard on the practical use of IC op amps. This book has earned respect in the industry by its comprehensive coverage of the practical uses of IC op amps, including design approaches and hundreds of working examples. The third edition has been updated to include the latest IC devices, such as chopper stabilized, drift-trimmed BIFETS. The section on instrumentation amps reflects the most recent advances in the field.
ISBN: 0-672-22453-4, $18.95

☐ Introduction to Digital Communications Switching   *John P. Ronayne*

Here is a detailed introduction to the concepts and principles of communications switching and communications transmission. This technically rigorous book explores the essential topics: pulse code modulation (PCM), error sources and prevention, digital exchanges, and control. Sweeping in its scope, it discusses the present realities of the digital network, with references to the Open Systems Interconnection model (OSI), and suggests the promising future uses of digital switching.
ISBN: 0-672-22498-4, $23.95

☐ John D. Lenk's Troubleshooting and Repair of Microprocessor-Based Equipment
John D. Lenk

Here are general procedures, techniques, and tips for troubleshooting equipment containing microprocessors from one of the foremost authors on electronics and troubleshooting. In this general reference title, Lenk offers a basic approach to troubleshooting that is replete with concrete examples related to specific equipment, including VCRs and compact disk players. He highlights test equipment and pays special attention to common problems encountered when troubleshooting microprocessor-based equipment.
ISBN: 0-672-22476-3, $21.95

☐ Microwave Oven Troubleshooting and Repair Guide *Jay R. Laws*

This complete but elementary reference for the layperson and electronics hobbyist assumes little knowledge of electronics. It features a detailed description of the major components of a microwave, how the components operate, and the common problems that can occur. It includes a useful section on preventive maintenance and cleaning. The more experienced technician will appreciate the section on advanced repairs and adjustments. For easy reference, a special appendix listing manufacturers and parts is included.
ISBN: 0-672-22481-X, $19.95

☐ Mobile Communications Design Fundamentals *William C. Y. Lee*

This authoritative introduction to mobile communications design, including cellular radio, provides communications engineers and engineering students with an interpretation of the mobile radio environment. It presents problems related to that environment and their solutions through the choice of properly designed parameters. Well supported by illustrations and examples, this incisive book explores propagation loss, calculation of fades and methods for reducing fades, interference, frequency plans, design parameters at the base station and mobile unit, and signaling and channel access.
ISBN: 0-672-22305-8, $34.95

☐ Computer-Aided Logic Design
Robert M. McDermott

An excellent reference for electronics engineers who use computers to develop and verify the operation of electronic designs. The author uses practical, everyday examples such as burglar alarms and traffic light controllers to explain both the theory and the technique of electronic design. CAD topics include common types of logic gates, logic minimization, sequential logic, counters, self-timed systems, and tri-state logic applications. Packed with practical information, this is a valuable source book for the growing CAD field.
ISBN: 0-672-22436-4, $25.95

☐ How to Read Schematics (4th Edition)
Donald E. Herrington

More than 100,000 copies in print! This update of a standard reference features expanded coverage of logic diagrams and a chapter on flowcharts. Beginning with a general discussion of electronic diagrams, the book systematically covers the various components that comprise a circuit. It explains logic symbols and their use in digital circuits, interprets sample schematics, analyzes the operation of a radio receiver, and explains the various kinds of logic gates. Review questions end each chapter.
ISBN: 0-672-22457-7, $14.95

☐ Landmobile and Marine Radio Technical Handbook *Edward M. Noll*

A complete atlas and study guide to two-way radio communication: private landmobile services, marine radiotelephone and radiotelegraph, marine navigation, and Citizens Band radio. Beginning with the fundamentals, this book covers everything from maintenance and installation to advanced systems and technology. It also discusses digital and microprocessor electronics, repeater stations and cellular radio, FCC licensing information, equipment testing and service, radar equipment, and satellite communications. An excellent text for radio communications courses or hobbyists.
ISBN: 0-672-22427-5, $24.95

☐ Principles of Digital Audio
Ken C. Pohlmann

Here's the one source that covers the entire spectrum of audio technology. Includes the compact disk, how it works, and how data is encoded on it. Illustrates how digital audio improves recording fidelity. Starting with the fundamentals of numbers, sampling, and quantizing, you'll get a look at a complete audio digitization system and its components. Gives a concise overview of storage mediums, digital data processing, digital/audio conversion, and output filtering. Filled with diagrams and formulas, this book explains digital audio thoroughly, yet in an easy-to-understand style.
ISBN: 0-672-22388-0, $19.95

☐ Radio Systems for Technicians
D. C. Green

This comprehensive volume examines the theory behind the broadcast and reception of radio frequencies. Discusses the principles of amplitude and frequency modulation; how to change the amplitude and frequency of a signal; how to build and use transmission lines and antennas; propagation of radio waves, RF power amplifiers; radio transmitters and receivers; and radio receiver circuits. An excellent reference for electronic technicians, radio enthusiasts, and hobbyists.
ISBN: 0-672-22464-X, $12.95

MORE
FROM
SAMS

☐ Electronics: Circuits and Systems
Swaminathan Madhu
Written specifically for engineers and scientists with non-electrical engineering degrees, this reference book promotes a basic understanding of electronic devices, circuits, and systems. The author highlights analog and digital systems, practical applications, signals, circuit devices, digital logic systems, and communications systems. In a concise, easy-to-understand style, he also provides completed examples, drill problems, and summary sheets containing formulas, graphics, and relationships. An invaluable self-study manual.
ISBN: 0-672-21984-0, $39.95

☐ Principles of Solid-State Power Conversion   *Ralph E. Tarter*
This comprehensive manual puts all the essential information about solid-state power conversion in an easy-to-use format. The author explores such fundamentals as vital protection and efficiency considerations, semiconductors, and component selection. He explains how to increase product efficiency and safety, improve performance, and achieve high-packing densities. Every facet of power conversion is detailed — switching systems, system operation and design, and passive devices. Over 20 years of hands-on experience is outlined in this excellent reference book.
ISBN: 0-672-22018-0, $44.95

☐ Reference Data for Engineers: Radio, Electronics, Computer, and Communications (7th Edition)
Edward C. Jordan, Editor-in-Chief
Previously a limited private edition, now an internationally accepted handbook for engineers. Includes over 1300 pages of data compiled by more than 70 engineers, scientists, educators and other eminent specialists in a wide range of disciplines. Presents information essential to engineers, covering such topics as: digital, analog, and optical communications; lasers; logic design; computer organization and programming, and computer communications networks. An indispensable reference tool for all technical professionals.
ISBN: 0-672-21563-2, $69.95

☐ Analog Electronics for Microcomputer Systems
Paul Goldsbrough, Trevor Lund, and John Rayner
This essential desktop reference covers power supply design, analog signal conditioning, data acquisition principles, data communication links, implementation of PID control, and other material necessary to microcomputer interfacing. Emphasizes basic theory, practical circuits, and the relationship of the peripheral analog circuits to the microcomputer.
ISBN: 0-672-21821-6, $19.95

☐ Howard W. Sams Crash Course in Microcomputers   *Louis E. Frenzel, Jr.*
Tired of wading through complicated books in computerese when what you really want is to learn quickly about microcomputers? Try this unique crash course approach. For those who need to know about microcomputers and programming, but do not have time for excess detail, this is your book. Learn BASIC programming, and enhance your understanding through photos, illustrations, and applications. Excellent for self-study or for beginning computer classes.
ISBN: 0-672-21985-9, $21.95

☐ 555 Timer Applications Sourcebook with Experiments   *Howard M. Berlin*
Describes the construction and use of various versions of the 555 timer and gives many practical applications.
ISBN: 0-672-21538-1, $9.50

☐ Semiconductor Device Technology
Malcom E. Goodge
This text explains fundamental principles of semiconductor technology, then discusses the practical operation and performance of commercial diodes, FETs, bipolar transistors, specialized switching, and optical devices. It shows in detail how planar fabrication takes place and thoroughly covers design, manufacture, and application of monolithic and film-type ICs. Contains tutorial questions with answers, information on network modeling, terminology, preferred component values, device numbering, and coding.
ISBN: 0-672-22074-1, $34.95

☐ Understanding Digital Logic Circuits
Robert G. Middleton
Designed for the service technician engaged in radio, television, or audio troubleshooting and repair, this book painlessly expands the technician's expertise into digital electronics. Beginning with digital logic diagrams, the reader is introduced to basic adders and subtracters, flip-flops, registers, and encoders and decoders. Memory types are also discussed in detail.
ISBN: 0-672-21867-4, $18.95

☐ Understanding IC Operational Amplifiers (2nd Edition) *Roger Melen and Harry Garland*
Technological advances are bringing us ever closer to the ideal op amp. This book describes that ideal op amp and takes up monolithic to integrated circuit op amp design. Linear and nonlinear applications are discussed, as are CMOS, BIMOS, and BIFET op amps.
ISBN: 0-672-21511-X, $9.95

☐ Transistor Fundamentals, Volume 2
Training and Retraining, Inc., Charles A. Pike
This introductory text explains transistor principles, voltage, current resistance, inductance, capacitance, and circuitry. It provides all information you'll need to develop a firm understanding of solid-state electronics and troubleshooting techniques.
ISBN: 0-672-20642-0, $9.95

☐ TTL Cookbook *Don Lancaster*
An early Lancaster effort that is still a tech classic. This is a complete look at TTL, including what it is, how it works, how it's interconnected, how it's powered, and how it's used in many practical applications. No technician's library is complete without it.
ISBN: 0-672-21035-5, $14.95

☐ Security Electronics (3rd Edition)
John E. Cunningham
This latest edition of a Sams classic introduces technicians and engineers to the fundamentals of electronic applications in the security area.
ISBN: 0-672-21953-0, $14.95

☐ Security Systems: Considerations, Layout, and Performance *William J. Cook, Jr.*
Design and install your own high quality home security system. Take a survey of your security needs, select the most effective sensing devices, lay out the wiring, and then complete the job yourself with the aid of this handy guide.
ISBN: 0-672-21949-2, $10.95

☐ Semiconductor General-Purpose Replacements (5th Edition)
Howard W. Sams Engineering Staff
Nobody knows replacement parts and cross-referencing like Sams. Our years of experience developing PHOTOFACT® service data have provided us information which is shared here. Shows general-purpose replacements for almost 225,000 bipolar and field-effect transistors, diodes, rectifiers, ICs, and more, listed by U.S. and foreign type number, manufacturer's part number, or other ID. Complete and easy to use.
ISBN: 0-672-22418-6, $10.95

☐ Practical Transformer Design Handbook *Eric Lowdon*
Here is transformer design from the ground up. The mathematics of transformer design and performance as well as concepts of electromagnetics are clearly presented. Separate chapters deal with general design considerations, transformer types, power losses, and the use of transformers in converters and inverters. A final section tells how to match off-the-shelf transformers to your custom requirements.
ISBN: 0-672-21657-4, $23.95

☐ Regulated Power Supplies (3rd Edition) *Irving M. Gottlieb*
Improved performance and greater reliability are the result when you use regulated power supplies. Learn about static characteristics, dynamic characteristics, regulation techniques, and linear and switching-type regulators using integrated circuits.
ISBN: 0-672-21808-9, $19.95

☐ RF Circuit Design *Christopher J. Bowick*
Enjoy the benefits of two books in one. Use this in cookbook fashion as a catalog of useful circuits or as a reference manual. It clearly presents a user-oriented approach to design of RF amplifiers and impedance matching networks and filters.
ISBN: 0-672-21868-2, $22.95

☐ The S-100 & Other Micro Buses (2nd Edition) *Poe and Goodwin*
This handy volume examines microcomputer bus systems in general, twenty-one of the most popular systems in particular, and shows how you can interface one with another.
ISBN: 0-672-21810-0, $9.95

MORE
FROM
SAMS

☐ **Microprocessor Circuits,**
Volumes 1 and 2   *Edward M. Noll*
Skilled microprocessor system installers, technicians, and designers are in great demand. These books can be your first steps toward understanding micro-controllers. Volume 1 presents basic microprocessor concepts and uses thirty progressive demonstration circuits to help you assemble, program, and operate a useful microcontroller system. Volume 2 covers I/O interfacing and programmable controllers in more detail, with another thirty demonstration circuits.
Volume 1
ISBN: 0-672-21877-1, $9.95
Volume 2
ISBN: 0-672-21977-8, $9.95

☐ **The Microprocessor Handbook**
Elmer C. Poe
This easy-to-use handbook contains complete, standardized specifications for today's most widely used 8- and 16-bit microprocessors. It includes data for popular parallel and serial I/O port chips and common memory chips. The specs for the 8080, 8085, Z80, 6800, 6802, 6809, 6502, 8086, 8088, Z8000, and 68000 are presented in a uniform format, allowing for easy comparison.
ISBN: 0-672-22013-X, $15.95

☐ **Modern Dictionary of Electronics**
(6th Edition)   *Rudolf F. Graf*
This comprehensive electronics dictionary clearly and accurately defines more than 23,000 technical terms dealing with computers, microelectronics, communications, semiconductors, and fiber optics. Over 3500 new entries and 5000 definitions, including abbreviations, cross-references, and acronyms, have been added, making this 6th edition the most up-to-date, all-inclusive electronics dictionary in the world.
ISBN: 0-672-22041-5, $39.95

☐ **Introduction to Automotive Solid-State Electronics**   *ASIA*
Understand the workings of selected on-board automotive "black boxes" and logic systems. Included are anti-skid braking, electronic ignition control, and trip computers.
ISBN: 0-672-21825-9, $9.95

☐ **Know Your Oscilloscope (4th Edition)**
Robert G. Middleton
The oscilloscope remains the principal diagnostic and repair tool for electronic technicians. This book provides practical data on the oscilloscope and its use in TV and radio alignment, frequency and phase measurements, amplifier testing and signal tracing, and digital equipment servicing. Additional material is provided on oscilloscope circuits and accessories. A vital reference for your workbench.
ISBN: 0-672-21742-2, $11.95

☐ **The Local Area Network Book**
E. G. Brooner
Localized computer networks are a versatile means of communication. In this book you'll learn how networks developed and what local networks can do; what's necessary in components, techniques, standards, and protocols; how some LAN products work and how real LANs operate; and how to plan a network from scratch.
ISBN: 0-672-22254-X, $7.95

☐ **Microprocessor Based Robotics**
Mark J. Robillard
Learn the mechanics of robotic arms and legs, tactile sensing, motion and attitude sensing, and vision systems. Through BASIC programs on a microprocessor, discover how to control the entire robot system by voice command. Well written and easily comprehended, this book brings tomorrow's technology to reality.
ISBN: 0-672-22050-4, $18.95

☐ **How to Build Speaker Enclosures**
Alexis Badmaieff and Don Davis
A practical guide to the whys and hows of constructing high quality, top performance speaker enclosures. A wooden box alone is not a speaker enclosure — size, baffling, sound insulation, speaker characteristics, and crossover points must all be carefully considered.
ISBN: 0-672-20520-3, $6.95

☐ The Howard W. Sams Crash Course in Digital Technology *Louis E. Frenzel, Jr.*

Back by popular demand, the "crash course" format is applied to digital technology. This concise volume provides a solid foundation in digital fundamentals, state of the art components, circuits, and techniques in the shortest possible time. It builds the specific knowledge and skills necessary to understand, build, test, and troubleshoot digital circuitry. No previous experience with digitals is necessary.
ISBN: 0-672-21845-3, $19.95

☐ IC Timer Cookbook (2nd Edition) *Walter G. Jung*

You can learn lots of ways to use the IC timer in this second edition which includes many new IC devices. Ready to use applications are presented in practical working circuits. All circuits and component relationships are clearly defined and documented.
ISBN: 0-672-21932-8, $17.95

☐ Electronic Telephone Projects *Anthony J. Caristi*

Perform fifteen fascinating telephone projects with the help of this book. Through building, testing, and connecting these projects, you will gain an understanding of basic telephone principles. The phone-user-to-phone-company relationship is also discussed.
ISBN: 0-672-21618-3, $8.95

☐ Fiber Optics Communications, Experiments, and Projects *Waldo T. Boyd*

Another Blacksburg tutorial teaching new technology through experimentation. This book teaches light beam communication fundamentals, introduces the simple electronic devices used, and shows how to participate in transmitting and receiving voice and music by means of light traveling along slender glass fibers.
ISBN: 0-672-21834-8, $15.95

☐ Guide to CMOS Basics, Circuits, and Experiments *Howard M. Berlin*

Why is complementary metal oxide semiconductor better than TTL? With this book you can learn what CMOS devices are, what are their characteristics, and what are their design rules. Twenty-two informative experiments which you can perform are included.
ISBN: 0-672-21654-X, $9.95

☐ Design of Phase-Locked Loop Circuits with Experiments *Howard M. Berlin*

Learn more about TTL and CMOS devices. This book contains a wide range of lab-type experiments which reinforce the textual introduction to the theory, design, and implementation of phase-locked loop circuits using these technologies.
ISBN: 0-672-21545-4, $12.95

☐ Digital Counter Handbook *Louis E. Frenzel, Jr.*

Here is a wealth of practical information on electronic and mechanical counter operation, specifications, and applications.
ISBN: 0-672-21758-9, $10.95

☐ Digital Logic Circuits: Test and Analysis *Robert G. Middleton*

No experience is necessary to learn digital circuitry with this book by performing basic digital tests and measurements as efficiently as possible.
ISBN: 0-672-21799-6, $16.95

☐ Don Lancaster's Micro Cookbook, Volume 2 *Don Lancaster*

More Lancaster advice helps you apply machine language fundamentals to the microprocessor family and microcomputer of your choice. Covers address space, addressing, and system architecture; machine code programming with the "those #$!&#%$! codes" techniques; details of input/output; and solutions for problems you may encounter.
ISBN: 0-672-21829-1, $16.95

☐ Electronic Music Circuits *Barry Klein*

Understand music synthesizers and how they work, then build your own. Each component in the synthesizer system is explained and illustrated. The components are ultimately combined into a do-it-yourself sample system with suggestions for modifications and enhancements.
ISBN: 0-672-21833-X, $16.95

☐ Electronic Prototype Construction *Stephen D. Kasten*

Breadboarding can be fun. Learn contemporary construction and design methods for building your working prototypes. Discusses IC-based and microcomputer-related schematics and ideas for evaluation and testing. Techniques include wirewrapping, designing, making, and using double-sided PC boards; fabricating enclosures, connectors, and wiring; and screen printing the panels, chassis, and PC boards.
ISBN: 0-672-21895-X, $17.95

☐ The Complete Guide to Security *Martin Clifford*

Complex electronic security systems are not just extensive, but also expensive. Elaborate measures are not required. This book helps you protect against burglary, theft, and armed robbery anywhere and provides a wealth of practical advice and information on simple self-protection devices.
ISBN: 0-672-21955-7, $13.95

MORE
FROM
SAMS

☐ Design of Op-Amp Circuits with Experiments   *Howard M. Berlin*
An experimental approach to the understanding of op amp circuits. Thirty-five experiments illustrate the design and operation of linear amplifiers, differentiators and converters, voltage and current converters, and active filters.
ISBN: 0-672-21537-3, $12.95

☐ Building and Installing Electronic Intrusion Alarms (3rd Edition)
John E. Cunningham
Protect yourself against robbery, burglary, and electronic eavesdropping! Learn the fundamentals of personnel identification and verification, as well as detection of concealed weapons.
ISBN: 0-672-21954-9, $10.95

☐ Analog Instrumentation Fundamentals
Vincent F. Leonard, Jr.
Probably the best book available on analog instrumentation. Subjects include ammeters, voltmeters, ohmmeters, bridges, filters, and attenuators. Hands-on experiments add clarity to the concepts presented.
ISBN: 0-672-21835-6, $19.95

☐ 555 Timer Applications Sourcebook with Experiments   *by Howard M. Berlin*
Describes the construction and use of various versions of the 555 timer and gives many practical applications.
ISBN: 0-672-21538-1, $9.50

☐ Basic Electricity and an Introduction to Electronics (3rd Edition)
Howard W. Sams Engineering Staff
Extensive two-color illustrations and frequent questions and answers enhance this introduction to electronics. The mathematics of electrical calculations are clearly presented, including Ohm's law, Watt's law, and Kirchhoff's laws. Other topics include cells and batteries, magnetism, alternating current, measurement and control, and electrical distribution.
ISBN: 0-672-20932-2, $11.95

☐ ABCs of Electronics (3rd Edition)
Farl J. Waters
A self-contained tutorial on the fundamentals of electronics. The many illustrations and review questions make this an excellent quick introduction to electronics concepts such as atoms and electrons, magnetic forces, and basic electronic components and their applications.
ISBN: 0-672-21507-1, $7.95

☐ Active-Filter Cookbook   *Don Lancaster*
Need an active filter, but don't want to take the time to design it? Don Lancaster presents a catalog of predesigned filters which he encourages you to borrow and adapt to your needs. The book teaches you how to construct high-pass, low-pass, and band-pass filters having Bessel, Chebyshev, or Butterworth response. It can also be used as a reference for analysis and synthesis techniques.
ISBN: 0-672-21168-8, $15.95

Look for these Sams Books at your local bookstore.

To order direct, call 800-428-SAMS or fill out the form below.

Please send me the books whose titles and numbers I have listed below.

Name *(please print)* _____

Address _____

City _____

State/Zip _____

Signature _____
(required for credit card purchases)

Enclosed is a check or money order for $ _____ (plus $2.00 postage and handling).

Charge my: ☐ VISA ☐ MasterCard

Account No. _____   Expiration Date _____

Mail to: Howard W. Sams & Co., Inc.
Dept. DM
4300 West 62nd Street
Indianapolis, IN 46268

DC039

SAMS ™